Mobile
User Experience

Patterns to Make Sense of it All

Adrian Mendoza

AMSTERDAM • BOSTON • HEIDELBERG • LONDON
NEW YORK • OXFORD • PARIS • SAN DIEGO
SAN FRANCISCO • SINGAPORE • SYDNEY • TOKYO

Morgan Kaufmann is an imprint of Elsevier

ELSEVIER

Acquiring Editor: *Meg Dunkerley*
Editorial Project Manager: *Heather Scherer*
Project Manager: *Punithavathy Govindaradjane*
Designer: *Russell Purdy*

Morgan Kaufmann is an imprint of Elsevier
225 Wyman Street, Waltham, MA 02451, USA

Library of Congress Cataloging-in-Publication Data
Mendoza, Adrian.
    Mobile user experience: patterns to make sense of it all / Adrian Mendoza.
        pages cm
    ISBN 978-0-12-409514-4
1. Mobile computing. 2. User interfaces (Computer systems)–Design. 3. Application software–Development. I. Title.
    QA76.59.M45 2014
    004–dc23   2013020414

British Library Cataloguing-in-Publication Data
A catalogue record for this book is available from the British Library

ISBN: 978-0-12-409514-4

Printed and bound in China
14 15 16 17 18   10 9 8 7 6 5 4 3 2 1

For information on all MK publications, visit our website at *www.mkp.com*

To my son Mateo, who teaches me new things every day, and who, at 8 months old, taught me that if you swipe the iPad screen with your entire hand it switches from one app to another. That four-finger swipe gesture took me 2 weeks to replicate.

This book is dedicated to you.

Contents

Chapter 6   Mobile UX Patterns   87

Acknowledgments

Thank you to my wife, Senofer, for not throwing me out of the house when I took on another large project, and for proofreading and editing this book. Thank you to Alyssa Arrigo and Brian Manning for reviewing pages and pages from each chapter. Thank you to Jason Grigsby for reviewing this book, even while on his vacation, and for the hours of phone conversations. Thank you to Ted Squire for teaching me to how to fly a spinnaker. Thank you to author Roger Warner for mentoring me throughout the process. Thank you to Steve Elliot and Meg Dunkerley from Morgan Kaufmann for pestering me to write this book until I finally caved in; I am glad I did!

Author Biography

Adrian's career is highlighted by more than twenty years of design and user experience work. His first studio, Synthesis3, worked with several Palm OS software companies in creating their brand for both a web and retail presence. His current business, Mendoza Design, specifically focuses on delivering user experience design for clients. Adrian has consulted on a variety of user experiences for web and mobile projects, ranging from orange juice to international airports. In addition, he is a cofounder of Marlin Mobile, whose focus is on creating tools to measure the performance of mobile user experiences.

Adrian has taught visual design at Suffolk University, Harvard University, The University of Southern California, and Massachusetts College of Art. His focus has been on creating a narrative using digital media. Adrian earned his BA with honors from the University of Southern California and his Master's from the Harvard Graduate School of Design.

Preface

A long time ago (I am talking more than 15 years here), I had the opportunity to meet the architect Wes Jones at a book signing for his second book, *Instrumental Form*, in Santa Monica, California. Among the buzz and giddiness of the architects in the room, I had an opportunity to talk with him in person. Unlike everyone else who asked him what his favorite building was or why his buildings looked funny, I asked him this one question: "What was the most memorable thing about writing your book?" He took a moment to think about this question. His thoughtful response was this: "After several months of collecting drawings and writing … after all this hard work. When all was said and done, this entire book and drawings fit in a tiny tape backup that could fit in the palm of my hand … this was the most amazing and memorable thing about writing this book."

This book was written on an iPad, two MacBooks, and an iPhone. It was written at my office, on a plane, at home, and in several libraries. For me there was no tape to hold in my hand to marvel at the compactness of it all. In fact, there was no CD-ROM, USB key, or portable hard drive. This entire endeavor was saved on cloud storage and shared to reviewers by email and file sharing. From here, I could open, edit, and save files virtually. By the time you read this book you probably will have downloaded it directly to your Kindle or iPad and waited only a few seconds for it to be copied over. This is the new memorable moment of this book. A book about mobile experience, created and read on a mobile device.

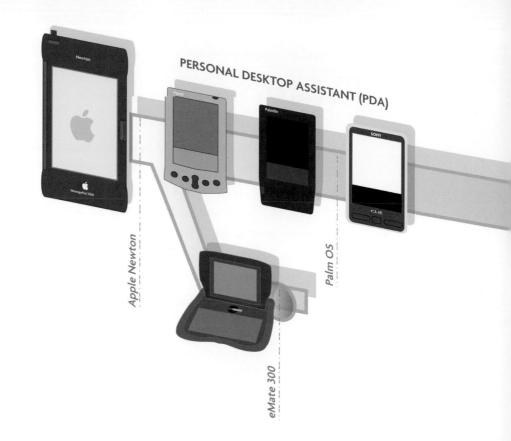

PERSONAL DESKTOP ASSISTANT (PDA)

Apple Newton

eMate 300

Palm OS

CHAPTER 1

Introduction

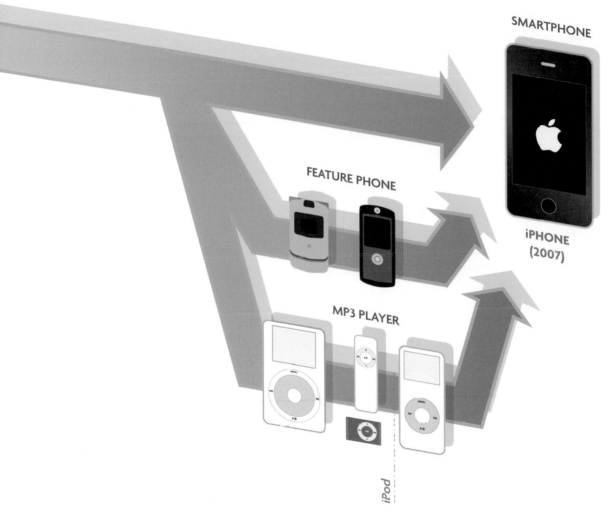

SMARTPHONE

FEATURE PHONE

iPHONE
(2007)

MP3 PLAYER

iPod

I have a confession. I once owned an Apple Newton MessagePad (Apple's first PDA), and along with that, a 14.4 dial-up modem ... Checking email on an Apple Newton was considered cool in 1998. Since then, I have owned an Apple eMate 300, a Palm V with a wireless modem, a Sony Clie, a Palm 3C, a Motorola RAZR, a Motorola SLVR, and a pile of different iPods, all leading to the iPhone 1G in 2007. You can track the history of mobile devices from the matrix of devices I have owned. As an experience designer, now specializing in mobile user experience, I have seen colossal changes in the user experience from that first Apple Newton to the Apple iPhone, almost 10 years later.

When I began working on user interfaces for Palm OS software in 1998, the form and experience of the user was secondary to the development and capability of the software, let alone the limitations of working with gray-scale displays. As feature phones[1] emerged, so did the ability to access the web on our phones, but only wireless application protocol (WAP)-enabled web pages were available. Imagine trying to view your favorite web pages, but only if a WAP version of the website was available. If you did manage to access the site with your limited Internet connection, you would see only text, and, if you were lucky, a few small images.[2] Designing a mobile experience was a challenging and almost futile effort given the maturity of the desktop web by the mid-2000s.

The iPhone arrived at just the right time and the right place. I don't want to say that it was luck, but the stars aligned to bring this kind of design and feature-forward device to an already engaged and hungry audience. By 2007, user experience design was no longer a luxury in web design but an integral part of the design and building process. Complement this with a huge market share of iPods and other emerging handheld devices (e.g., BlackBerrys, GPS devices) all with highly evolved user experiences; we had both consumers ready to consume the web on mobile devices and designers ready on the other end to design interfaces for them. The mobile age was upon us, and it began by a touch and a slide.

For the first time, mobile devices impacted the rest of the computer industry; user experience design and the entire software industry, was turned on its head by something that could fit in your pocket. The iPhone delivered experience designers

---

[1] Feature Phone—Term for mobile phone with basic Internet connection. It mainly opens WAP enabled web pages.

[2] Stan Miastkowski, "How It Works: Wireless Application Protocol," PCWord.com, July 3, 2000, http://www.pcworld.com/article/17437/article.html.

a platform focused on user interface (UI) and user experience (UX) design; Apple first published its own UI/UX guidelines,[3] which were available when the first iOS SDK was released. As other mobile platforms followed the iPhone, and Android and Windows emerged, they also provided guidelines to help build and design for their own experiences.

As user experience designers, we now had a new challenge of migrating our web experiences to the new "mobile web" and "mobile apps," two platforms that emerged simultaneously. I can still remember one of my first mobile app projects from early in 2009. Our client's only direction …"Make the app fun!" This wasn't a typical request, and coming from a client in the financial services realm, maybe a little strange. At that point, the mobile was still the uncharted wild, wild west, and apps were a novelty.

The more I worked on mobile projects, the more I saw the gap widening between the design of user experiences in mobiles compared to those on the desktop. Designers of mobile experience had to take into account new interfaces, new hardware, unique experience design to each mobile platform, and keep those designs updated to include the changing form factors of devices. Mobile user experience was truly a unique discipline that needed its own language for describing, visualizing, and creating experiences. Before Apple released their standards, no standards existed for terminology and graphics to help describe mobile design.

The idea for this book took a few years to evolve into a cohesive notion. I wanted to help answer the questions: *How does one start learning how to design for mobile UX? What are the differences compared to the desktop? And what makes mobile UX unique?*

This book is a call to action for creating, learning, and visualizing mobile experiences in a structured, effective manner. The outline and direction are simple: Use a series of common user experience patterns to describe and implement mobile

---

[3] Originally called iPhone Human Interface Guidelines.

experience. These are a set of conceptual patterns that can work across platforms, device, and users. Yet to balance the conceptual, we need to define the practical methodologies and tools to demonstrate the experience. *How do we wireframe? What are the user interface elements and code elements to create them, and how do we distinguish them across the unique platforms? How do I prototype those experiences?* My goal with this book is to create the mobile UX guide/roadmap, to empower the user experience designer, inform the developer, and engage the curious reader on mobile user experience design. Our industry will shape how people interact with mobile devices, and this book is meant to give us the tools and languages to communicate that experience.

This is not the manifesto of a sole designer; it is one that we will build together.

Now, let's build some mobile user experiences!

CHAPTER 2

Why Mobile UX, Why Now?

INTRODUCTION

A few years ago, I lectured at a college about the future of the mobile web.[1] At the end of the lecture, one of the faculty members approached me with a question. "In the battle between print and the web, who will win?" Taken aback, I answered with: "The war has been over for a while, now we are just counting the bodies ..." Now imagine what I would have said had she asked me about the mobile.

In the case of the mobile, it was not a battle but a leveling impact with the force of an atomic blast. Its effects were fast and far-reaching. The mobile went from being a

---

[1] Adrian Mendoza, "Design in the Age of Mobility," MassArt, Boston, MA, October 28, 2010.

toy one day to a source of revenue the next. By early 2011, smartphones had outsold desktops.[2] By 2013, worldwide smartphone sales were in the billions (1.8 billion to be exact).[3] To put this in perspective, with the "personal desktop revolution" that started in the mid-1980s, it has taken more than 20 years to get worldwide PC sales over 300 million (2005).[4] With this trend, some users, especially in developing countries, will access the web on a smartphone before they do on a desktop; this is already the case in some countries. We have entered the mobile age and it is coming straight at us like a storm, *a perfect storm*.

THE PERFECT STORM OF MOBILE

Why a perfect storm? Like the figure of speech for the weather, this storm "arose from a rare combination of adverse meteorological factors."[5] It took three unique factors to align and develop to create the mobile storm we are in now. We needed the technology, the users, and user experience all ready at the same time. Had one or another not been available, the ecosystem would not have evolved. Let me explain.

TECHNOLOGY

The first factor is the technology itself. By the time the first iPhone appeared, small devices were already commonplace. Users already had a history with iPods, mp3 players, and PDAs; the general public were accustomed to small handheld devices. Feature phones[6] used some form of Internet access and these were common across the globe. By this time everyone, including his or her brother, grandmother, or child

---

[2] Dana Wollman, "Smartphones Outsell PCs for the First Time Ever," *Huffington Post*, February 02, 2011. http://www.huffingtonpost.com/2011/02/08/smartphones-outsell-pcs-f_n_820454.html.

[3] Gartner.com, April 4, 2013, http://www.gartner.com/newsroom/id/2408515.

[4] eTForecasts, "Worldwide PC Market." http://www.etforecasts.com/products/ES_pcww1203.htm.

[5] Oxford English Dictionary.

[6] Feature Phone—Term for mobile phone with basic Internet connection. It mainly opens WAP enabled web pages.

had a Motorola Razor phone in their pocket. It was the inclusion of touch technology and readily available 2G bandwidth that accelerated the use of the smartphone. The synthesis of a small handheld device with a phone, mp3 player, and camera was the next logical step in the technology trend.

USERS

The second factor is the users themselves. Like the technology, they were already familiar with keeping a small handheld device in their pockets. They were familiar with of using and carrying an mp3 player, phone, and/or camera (sometimes carrying all three). Combining these three devices was the logical next step. The groundwork for the usage patterns already existed, and adding Internet usage was a natural expansion. By the time the first smartphones were launched in 2007, users were already familiar with using the desktop web. There was no need to teach them about buying online, accessing their email, looking for a map location, and other web-based patterns. The learning gap and usage patterns of getting users online had already been bridged. Some users had been using the web since primary school at this point, users were not only well versed, they had grown up with the Internet. The only thing left was to teach them how to optimize these patterns for mobiles; a smaller feat compared to teaching them how to use the web on a desktop computer for the first time.

USER EXPERIENCE

The third factor of the perfect storm is user experience as a discipline. By 2007, user experience design was already considered a formal discipline in the web design industry. It had become a crucial piece of any web-based design or build. It has evolved from the art form in its infancy (information design) to the science of user centric experience design. Now a user experience designer sits between an art director and the web developer; at times they even manage the project. The science of user experience design is also the guiding principle of increasing web sales, increasing user conversions, optimizing design, and retaining users. Web design has been refined from a loosely trained discipline (you usually picked it up on the job) to the formal education of user experience designer (college degree and certificates). By the time Apple released the human interface guidelines for their iPhone, it was not a surprise, but rather an expectation and common practice.

THE PROBLEM

Like learning how to use the web on the desktop for the first time, mobile brought its own challenges. Compared to the desktop, we have introduced new variables to creating a user experience on a mobile device. Gone are the consistent factors of the desktop experience, the large size of bandwidth, and even the user's focus and attention to the task. With mobile, you have to take into account the carrier, the type of device you are on, the operating system, and the screen size; these factors add up to what I call the "Mobile Equation."

Carrier × Device × Operating system × Screen size = Your mobile user experience

The Mobile Equation.

Think of this as the recipe for designing mobile experiences. Mix the inconsistency of the carrier and your signal strength depending on your location. Stir in the different operating systems (Android, iOS, Windows, etc.) and the different versions of each OS. Whisk in the different form factors and sizes of each mobile device you will be using. Finally, add in the different screen sizes and types into the equation (LCD, Super AMOLD displays) and the unpredictable oven temperature of the users' usage patterns. This entire equation gives you an insight into the unique challenges of creating a mobile experience. Why would anyone want to inherit this mess? Why would anyone want to work on mobiles compared to the desktop? These are questions that can only be answered by the opportunity that the mobile provides us.

THE OPPORTUNITY

Regardless of its rapid growth, innovative technology, or even global reach, mobile gives us an opportunity that we have never had with the design of desktop web experiences. The grand vision of Web 2.0[7] was to connect one-on-one with a user; it was to provide them with personalized content and a user experience just for them through the use of dynamic web experiences. In reality, that dream never came to fruition; the web experience became too busy and overcomplicated the longer we went on. The mobile enables that dream to come true. It puts one Internet-enabled device in the hand of a single user. It gives them the power to access the web at anytime and at any place and the option to broadcast or connect directly one-on-one. It is this usage pattern that presents us the opportunity to design a mobile user experience that can be consumed on a one-on-one basis. Users will carry our mobile experiences with them anywhere and open them anytime. It is this same mobile experience that they can engage with and keep in their pocket; it is both their personal and public persona. The mobile gives a single user a camera, an open web connection, a way of posting, tweeting, and messaging to the outside world. It is their voice; you can help design a mobile user experience to enhance it. *This is the mobile opportunity*.

---

[7] Web 2.0— Term coined in early 2000s to describe dynamic websites with user generated or social media content.

CHAPTER 3

Your Desktop Experience Is *Not* Your Mobile Experience

INTRODUCTION

Like working in any medium (painting, sculpture, or even user experience), we need to learn a set of basic rules that act as guiding principles for our craft. For the mobile, I refer to these as *mobile mantras*. Like any mantra, these should be repeated, shared, and—if you want to—chanted. You might consider them practical, conceptual, or even spiritual; regardless, these mantras will help guide your journey when creating a mobile user experience. Let's start.

> **MOBILE MANTRA #1:**
> YOUR DESKTOP EXPERIENCE IS *NOT* YOUR MOBILE EXPERIENCE

The mobile experience is a completely separate entity from the desktop experience. If everything about the mobile is different—its screen size, Internet connection, device size, and the mindset of the person as they approach the site—why would you take your existing desktop experience and copy it? This would be like trying to fit 3 pounds of sugar in a 1-pound bag. I would even go as far as pushing the simile of a mobile from a 1-pound bag to a 1-ounce bag. This may sound a little overly dramatic, but in fact the differences between the desktop and mobile experiences are vast and provide the two with distinct strengths and weaknesses. Let's take a look at a few.

MOBILE USE IS *NOT* DESKTOP USE

In reality, we spend too much time in front of our desktops; we are searching and researching and, when we are done, we search again. As a result, we create larger user experiences catering to the longer timeframe that users focus on a page. When taking time to use a desktop, users are usually at home, in a café, on a plane, or somewhere where they are seated and taking a few moments to open and boot up their computer. They are focused on the screen. When looking at the mobile, we are looking at smaller increments of time that a user spends in front of their smartphones. An average time per interaction on a mobile device is 17 minutes compared to 39 minutes on a desktop.[1] When people are on their phones, they are in a conversation, commuting, or waiting for a meeting; they are interacting with the phone in another environment not centered on the phone. As a result, we need to take into account this bite-sized experience and design for it.

MOBILE USERS ARE *NOT* YOUR DESKTOP USERS

Remember the great marketing debate between Apple and PC users? Each seems so different. In reality, the differences are miniscule compared to that of desktop versus mobile users. The difference between mobile users themselves is at times staggering; iPhone users are younger and more affluent (higher income bracket), Android users view more mobile content (while iPhone users are more engaged with the content), and iPhone users are more likely to make purchases on a mobile device than Android users.[2] This is just a sampling of trends. Add into the mix your own customers and

---

[1] Google, "The New Multi-screen World: Understanding Cross-Platform Consumer Behavior," August 2012.

[2] Comscore, "Android vs. iOS: User Differences Every Developer Should Know," March 6, 2013, http://www.comscore.com/Insights/Blog/Android_vs_iOS_User_Differences_Every_Developer_Should_Know#imageview/0/.

users, and you now have a more complex view of the mobile user ecosystem. Feels like that's not enough? Add into the mix the differences between smartphone and tablet users, you have an even more complex view of your mobile users. Take all of this into account and include different devices, carriers, operating systems, and screen sizes (the Mobile Equation); your mobile users look more and more different from their desktop counterparts.

YOU CANNOT FIT EVERYTHING ON ONE SCREEN

Having desktop screens that are now 1200 pixels wide and growing has spoiled us. The typical web experience uses several columns of information and pages to organize its content and functionality. Good user experience for the desktop tells us to lay out all of the options as a large sitemap and use wizards to walk us through a step-by-step experience—the more options available on the page the better. The mobile takes this away and creates a single, linear story. The more screens a user has to scroll through means death for our mobile experience. Cramming everything on a screen and trying to make it readable is now replaced by delivering only the bare essentials, one easily digestible piece at a time.

CUSTOMERS DETERMINE YOUR USER MOBILE EXPERIENCE

The role of customer feedback is more important than ever. A mobile customer will do less than your desktop customer. Finding out what they will be able to do is critical to building your mobile experience. Feedback and customer conversations play a critical role in designing your mobile strategy. What will they be willing to do on a tablet versus a smartphone? What do they want to do on a mobile? Does that compare with what you want them to do? Getting these answers will be the first step in creating your mobile experience. Remember that by not doing so, your phone number is just one touch away … and so is their complaint. Every complaint from an unhappy mobile customer to your call center will cost you money and brand equity.

RETHINK HYPERLINKING

Hyperlinking has always been the unique characteristic of the constant desktop web surfer. If you want to go somewhere new, it is just a click away. And while you are at it, open a few tabs, windows, or different browsers at the same time and toggle between all of them. This is a less fluid experience in a mobile. By necessity, the mobile provides a streamlined, linear experience. Going back to our point about knowing your mobile web user, you need to have a clear idea of what this path should be in order to construct it.

Our mobile browser is much smaller and more difficult to navigate. The concept of tabs on the mobile browser has been replaced with browser windows that open on another screen; the user experience of moving back and forth from one window to another is difficult, breaking the pattern of multi-window web surfing. If users are forced to open a new browser window, they are unlikely to go back. If a screen is too crowded, they are unlikely to take the time to sift through it to find what they want. The default browser loading into our devices has also limited unique user experience flows. Until recently, we have seen the rise of other browsers as apps and newer browser engines (Chrome app, Firefox app, Blink browser engine); all of which have their own user experiences and compatibility issues. This adds another layer of inconsistency for surfing our mobile experience.

Working on an app instead? As your workflow is limited to being only inside your app, you will need to rethink how to capture the user's attention and focus; if not, one push of a home button and they are out of your mobile experience.

THE MOBILE BROWSER IS LESS FORGIVING

One way or another something in your mobile experience will be connected to the Internet or web browser. Most mobile apps use what's called the *embedded web view* to open a mobile web window from inside the app; this is a very popular trend in app design. Gone are the days where we can launch websites that don't have cross-browser compatibility. Impressed by your fancy parallax effect[3] on your new website? Look at it again on a mobile browser; it more than likely won't be working. Impressed by all the fancy AJAX script you have on your page? Be prepared to faint when it takes over 10 seconds to load. If you don't think you are mobile, think again. Users are already accessing your web pages on mobile devices. Hello websites, welcome to the mobile.

SAY GOODBYE TO THE MOUSE

I will be the first one to write the obituary when we finally bury our mice. For many years we have used the mouse as a crutch. By removing the pinpoint accuracy of the mouse with touch gestures, we are forced to rethink the user experience for inputting information. Without the pinpoint accuracy of a mouse, forms and complicated selections need to be completely redesigned. As a result, it has focused us on making smaller pages, cleaner information layouts, and larger user interface elements. We will have to optimize user experiences for mobiles, making them better for our users. Goodbye mouse, hello touch.

BUILDING A MOBILE EXPERIENCE IS MORE COMPLICATED

Unlike working on the desktop, building an app or mobile website requires user experience design, user interface design, and the development to be completely integrated from the start of the project. Want to change the UI of a button? Well, by doing so you have just changed the user experience pattern and development time. If you decide to change the experience pattern for a screen, you have just added more work for the user interface designer and developer. If you decide to do a quick change in your iOS and Android code, you have possibly changed the user experience and UI design across multiple screens or for multiple devices. Gone are the days of fly-by-night web development, where anyone with a text editor could change a web page. What has arrived is the integrated and interconnected world of mobile design and development.

---

[3] Parallax Effect—Scrolling effect on desktop websites that overlaps images.

Finally we get to create great experiences as they are meant to be! But we do need to do a bit of planning to get there.

PLANNING YOUR MOBILE EXPERIENCE

The next step in preparing a mobile experience is to create a plan. Implementing a user flow, we can explore the differences between our desktop and mobile scenarios. A user flow allows us to lay out a preliminary schematic design of how our users will explore our experience before we get to wireframes. You might create different user flows for a desktop, iPhone, or Android experience; one design will not fit all. Implement a user flow to plan out specific mobile functionality.

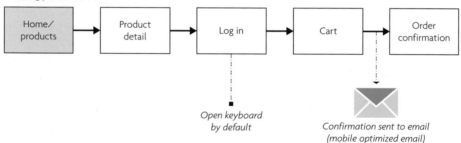

Example of Mobile User Flow #1.

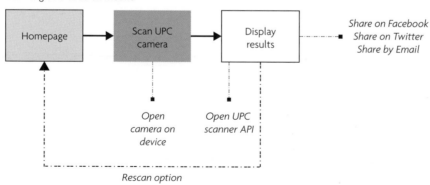

Example of Mobile User Flow #2.

Mobile User Experience | Patterns to Make Sense of it All

Depending on the mobile experience you want to create; the user flow needs to answer a few important questions:

1. Is it small?—Show that the path is short enough for a mobile user to complete.

2. Is it optimized for mobile?—Show any use of the camera, accelerometers, or other mobile-specific features.

3. Is it optimized for our mobile users?—Show that the path is clear for what the mobile users want and expect; make it easy to complete this path or function.

4. Is it designed for a specific platform?—Create a flow tailored to the mobile specific platform that you are designing for.

5. Does it integrate the *embedded web view* if it's an app?—Show areas in your app where you will display mobile web content.

Once you complete your plan, you can compare it with your desktop strategy. The mobile may have its own separate and unique strategy, but it does not mean that it cannot complement or supplement your desktop experience.

LET'S MOVE FORWARD

Sometimes you need to take two steps back to take one leap forward. Great user experiences have a tendency of being copied and reused. A good user experience you make today may show up in another application tomorrow. Don't be afraid to look, learn, and experiment. Like my example of the mouse, mobile experiences are already having great influence on other mobile platforms and the desktop. Even if you have never designed or built for the desktop, your mobile experience may one day help influence a desktop experience. The differences in the desktop and mobile we see today might not be the same in a few years or even a few months. Remember that a growing number of users may never touch the desktop at all and at a minimum will visit your mobile experience first. Your mobile website will be their entire introduction to the brand, website, and customer experience you are trying to create. And we all know the importance of first impressions.

CHAPTER 4

Understanding the Device

INTRODUCTION

> **MOBILE MANTRA #2:**
> BE ON THE DEVICE

The first time I heard this, I laughed at the phrase. It was such a ludicrous concept. Why would I need to be on the device? Why would someone own every phone model? This wasn't the case on my Palm OS device or Apple Newton, but the smartphone was a different beast. The differences between the first iPhone and the first Android device were night and day. Not only was the hardware distinctive, but the user interface alone was enough to transform my opinion about this phrase.

Now, after spending time with smartphones, I am a big believer of "being on the device." It has become a mantra I teach all designers and developers I meet. It is the first phrase I share when I am asked about the differences between mobile platforms and the last thing I repeat to others when I am asked to test an application or mobile website. The mantra is rooted in a single premise: to best understand an iPhone or Android device you need to feel it in your hand, to experience it personally. Only in this way can you capture the tactile and complementary nature of how interactions affect the UI. Only in this way can you answer the questions of differences between any smartphone. Without the experience of actually using what you have designed on a device, you have no way of refining it. How does my hand reach for a button? How big do my buttons and UI elements need to be when I touch or swipe? What happens when I first turn on the device? What happens when I open or close an app or mobile website?

If you are planning on designing or developing in iOS, get an iPhone; if you are planning on Android, get an Android device—or two or three. If you are planning on working on a tablet, get a tablet…. You can see where I am going with this. Only by being on the device can you get the true user experience and a thorough understanding of how you can shape the experience to account for difference between devices, operating systems, screen sizes, and form factors.

The goal of this chapter is to provide a primer for starting to design on multiple platforms. To begin designing for each experience you need to know how the platforms work and what are the unique user interactions, user interface elements, terminology, and device characteristics. Once you have the basics you can start building your unique experience across each platform.

GETTING TO KNOW iOS
AN ANECDOTE ABOUT GESTURES

Microsoft Surface Table Circa 2007.

Hidden away in the lobby of one of the big digital ad agencies in Boston sits a piece of mobile history. Cluttered with some fancy modernist seats, a digital vending machine, and a slick over-designed reception desk sits a table that you might confuse for a vintage Pac Man tabletop game if you didn't give it a second look. In 2007, Microsoft released their first version of a touch-based operating system called *Surface*. Integrated into a tabletop, most passersby in this lobby would not even know that this table was the one of the first touch-based interactive experiences.

In 2010 my two business partners and I sat in this company's lobby, waiting for a meeting. I walked up to the table and swiped to unlock it, selecting an image with both hands, I began to scale and rotate images and applications. With a flick of a finger, a push of my hand, I began to control the user interface like a pro. With a puzzled look, my two partners stared in disbelief—How could I learn and control all of these functions of this operating system in such a short time? One of them finally asked, "Have you used one of these tables before?"

My response: "No I have never used one of these before in my life." To their surprise they both asked "How did you learn how to use it so fast?"

"YouTube."

INTRODUCTION TO BEHAVIORS AND GESTURES IN iOS

EXERCISE #2:
AN INTRODUCTION TO THE iPHONE

Ask your friends, family, and colleagues this one simple question: What was the first thing you did when you got your iPhone?

To your surprise you will get back several responses ranging from: I plugged it in, I called my mom, I opened the box, I synced it with iTunes, I started surfing the web, downloaded an app … and a few others that you might not expect.

Mobile User Experience | Patterns to Make Sense of it All

In reality, all iPhone owners will encounter the same first learning experience without even realizing it. They have all encountered the welcome screen as the iPhone turns on for the first time.

In less than a second all new iPhone users learn all of the basic interactions in using the device:

1. Touch the home button with your finger to turn the phone on.

2. Touch and slide the screen to unlock it.

As a user experience designer, this is the pinnacle of what we refer to as the "5 second learn." In these critical first five seconds of grasping a user's attention the iPhone has taught its new users how to use it. In this basic screen the mobile experience has also taught the users some basic interactions as well:

1. *The screen is touch based.*

2. *User interface elements are touch based.*

3. *The user will need to use fluid gestures, such as touch and swipe to engage the user interface elements.*

4. *The hardware buttons are secondary to the touch experience.*

There is no manual with required reading, no long description for the users to read on the screen, no special alphabet to learn, no tutorials to go through, no call to customer support and definitely, no YouTube videos to watch. In less than 5 s, the user has been introduced to the world of iOS and using its touch gestures.

Behaviors of iOS

iPhone Home Screen.

Gestures were not the only innovative concept presented with the first iPhone. Once past the welcome screen, the user encounters a home screen with icons. These icons are designed to launch applications from this landing page. This concept of the "App" was new to users with the first iPhone release. Its use of icons as navigation and their relationship back to the "App" was critical in creating a scaled-down version of the desktop experience. This mobile experience would not mimic its desktop counterpart, but bring familiar desktop processes to people in more "bite-sized" chunks; hence the "app" was born.

The idea of an icon-based launcher was nothing new to mobile audiences; this concept had been previously seen in the Palm OS world. Unlike it predecessor, the iPhone added a section to the bottom of the screen to launch primary applications and a phone dialer. This idea of a bottom-based navigation would be a consistent navigation element that would be commonly deployed across most platforms. Another and perhaps the most lasting and influential innovation was Apple's creation of a standardized library of intuitive gestures, interactions, and user interface elements to produce a consistent user experience.

The Language of Touch Gestures

The iPhone introduced the world to a series of unique touch gestures. The idea of controlling a digital interface with a touch was not new; several years earlier the ability to touch and activate a screen was very common in kiosks and game devices. What made this event new was the language of gestures that was introduced in the iOS to complement its initial touch gesture. These gestures have since become part of the basic lexicon of how people approach devices.

The Tap

The tap gesture is the building block of the iOS platform. The uniqueness of this element is the usage of pressure sensitive glass to enable the finger pressure of the user to come in contact with the screen, a feature now universal to all touch-based mobile devices. This gesture engages the user interface elements using a minimum 44 × 44 point active area around a button or slider. This accounts for the signature larger, rounded UI elements that the iOS is known for. The pressure sensitive screen allows the user to hold down as lightly or as forcibly as needed. This use of sensitivity allows for tapping, flicking, and dragging of a finger using the same touch gesture, it is the building block of the gestures explained below.

Tap Gesture.

The Drag

Dragging uses the touch gesture to scroll or pan a user interface element. This gesture combines a push and movement of a finger to move an on-screen element. It is usually reserved for a steady and directed movement on the screen. An example of this would be to drag and move an app icon, or in the example of the welcome screen to drag the slider over to open the phone.

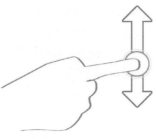

Drag Gesture.

The Flick

A user can use a flick to scroll or pan quickly through the screen or navigation elements. This gesture, like that of the drag, uses both the touch and movement of a user's fingers. Unlike the drag, the flick is designed to allow for a lighter touch for quicker and less directed movement. The flick uses forward motion of a finger to start the gesture. An example of this gesture is scrolling to the bottom of a web page quickly or flipping through photos.

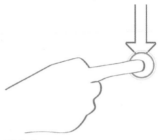

Flick Gesture.

The Swipe

The swipe is based on using a larger finger contact area for directed on-screen movement (famous for flipping through photos in iOS). The swipe allows the user to access on-screen menus, access navigation trays, open menus, and other touch-heavy user interface elements. A common example on the iPhone is to swipe from one home screen panel to the next, moving from one element to another in the iTunes carousel, or accessing the dropdown notification menu. It is meant for slower and more controlled interactions.

Swipe Gesture.

The Pinch

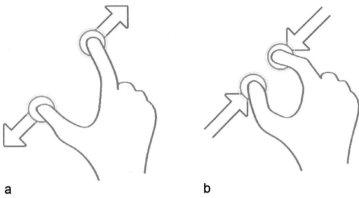

a
b

a. Pinch Gesture—Pinch out (Zoom in), b. Pinch in (Zoom out).

The pinch uses finger interaction to control zooming in and out of an application or screen. It is based on two distinct gestures. The first is the use of a pinch open to zoom in and the second is the use of a pinch close to zoom out. This is the first iOS gesture to use two contact points to activate the on-screen controls. This gesture allows users not only to zoom in and out of apps, like Google Maps, but also to zoom in and out of web pages.

> TOOL TIP
>
> Want to learn more about other touch gestures and those available across different platforms? I have collected a reference list with various links and articles. View it here:
>
> http://www.mobileuxbook.com/gestures.

Random Gestures

Along with these basics, some other gestures have been introduced that users are less familiar with. One of the more random gestures introduced in iOS is the "Shake." It uses the phone's accelerometer to activate a user interface element or user-driven action. Not commonly known, one of the only examples found in iOS is using the Shake to undo a typed word or to provide feedback in Google Maps. If you ever see a person randomly shaking their iPhone like a crazy person, don't be alarmed, they are using the shake gesture. …

No really.

THE iOS KIT OF PARTS

Apple's iOS Human Interface Guidelines focus on creating a standardized user experience for its hardware and user interface use. I would not call these guidelines but requirements. Designing to these specifications, interactions, and user interface elements is necessary when applying for their app review process; a process required when publishing an app in the Apple App Store.

This is not to say that this process is difficult or impossible, but rather the focus of these guidelines is to create a consistent user experience. By introducing you to the iOS kit of parts in this chapter it can help you understand these "guidelines" and plan for them when you design your mobile experience.

Screen Sizes

a. 320 × 480, b. 640 × 960, c. 640 × 1136.

One of Apple's first guidelines was standard screen sizes. This way a designer and developer could build to a set screen size and resolution. At this moment in time, there are three iOS screen sizes. Regardless of whether you design an app or a mobile website, you must be aware that you will need to take into account these three resolutions.

Standard iOS Display—320 × 480

Used for the iPhone 1G (first iPhone) to the iPhone 3Gs, this screen size became the standard size for most mobile devices. The screen size is now obsolete but it still exists on some phones. Consider it the minimum screen size to design for.

Retina iOS Display for 3.5" Screen—640 × 960

A new entry for Apple iPhone 4 and 4s only devices, the retina screen is Apple's attempt to introduce a high-resolution display. The screen is physically the same size as the old 320 × 480 resolution. Most users will not notice the difference when you put the two phones next to each other. As a designer you will need to design for the larger screen resolution. Apple also introduced the "@2x" file format; this tells the Safari browser (with the help of CSS and Javascript) and Apps to display images at a high resolution. This format was introduced to iPad and iTouch devices as well.

Retina iOS Display for 4" Screen—640 × 1136

Introduced for the iPhone 5 and above, the screen size also retains the same physical screen as the first retina display … just a bit longer … by ½ inch.

User Interface Elements

The core interface elements in iOS are referred to as "standard controls." This library of buttons, navigation, keyboards, and other common entry points to interaction maintain their own look and feel with some basic options for customizing them. When building and designing an app, the color palette of these UI elements can be changed, but the basic shape and function will always remain the same. These elements also have built-in hover and on-press states. To change the shape or function will require additional design and development effort. This is what is referred to as building a "custom control."

When designing for the mobile web the input controls (i.e., keyboard, spinners, and pickers) become standardized by the browser. For example, if you have a field in a web page, the browser will open whatever default keyboard iOS is using. By using HTML 5 some inputs can be specified. In these cases I will add the code for you to try.

The Keyboard

Example of iPhone Keyboards.

The iOS keyboard has matured since its first release. Now the keyboard input includes not just alphanumeric values, but language localization, emoticons, and special characters. One of its signature UI elements is an over state triggered when a finger touches the button. This function also allows a 2s touch to access special characters per each letter. As a designer, this gives you a fully stocked keyboard to work with; best of all, the proportion and size of this keyboard remains consistent through OS versions and different device models (except in an iPhone 5, where the keyboard loads a ½ inch lower). This element, regardless of screen size, will always launch fixed to the bottom of the screen.

iPhone Screen Sizes with Keyboards.

When designing for an app you have the unique function in iOS to specify which keyboard you want to launch. The keyboard types include:

Default

Email

URL

Phone

iPhone Keyboard Types.

HTML 5 CODE

By using HTML 5 on your mobile website you can also access these keyboard options. Add this code to your input fields. Try it out!

- Open a Text Keyboard: `<input type="text"></input>`

- Opens a Number Keyboard: `<input type="number"></input>`

- Open a Telephone Keypad: `<input type="tel"></input>`

- Open a URL Keyboard: `<input type="url"></input>`

- Open an Email Keyboard: `<input type="email"></input>`

- Open an Zip Code Keypad: `<input type="text" pattern="[0-9]*"></input>`

- Open a Date Input: `<input type="date"></input>`

- Open a Time Input: `<input type="time"></input>`

- Open a Date & Time Input: `<input type="datetime"></input>`

- Open a Month Input: `<input type="month"></input>`

The Pickers and Date Pickers

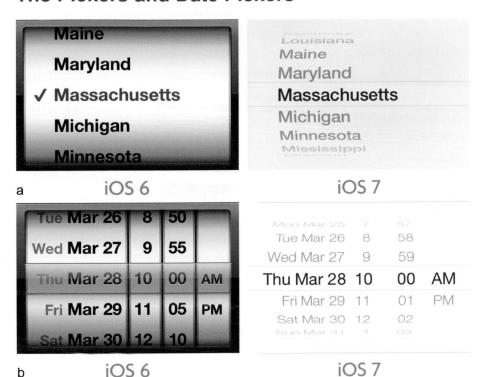

a   iOS 6   iOS 7

b   iOS 6   iOS 7

a. Picker, b. Date Picker.

The picker is another unique iOS element. The picker replaces the traditional dropdown box from the desktop browser. This element allows for a larger contact area for touch gestures making it easier to pick from the selection options. When designing for the mobile web this option is loaded in the browser by default. On the other hand, when working with an app, you have the ability to add a date picker to your experience design. This is a combination of various inputs in one spinner; this input method can be customized to select from whatever categories you want. A good use of this UI element would be selecting currency or date and time. The pickers, like the keyboards, launch fixed to the bottom of the screen.

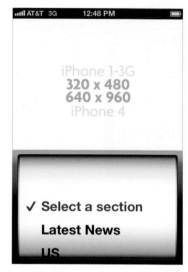

iPhone Screen Size with Picker.

Inputs

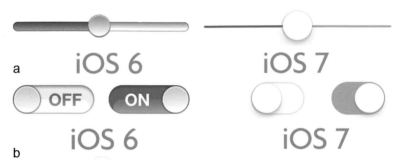

a

iOS 6

iOS 7

b

iOS 6

iOS 7

a. Slider, b. Switches.

The last set of inputs is app specific; they are the slider and the switch. These input methods can be used to create interactive methods when selecting choices and ranges on the screen. Their design emphasizes using gestures to control elements that would have normally been radio buttons or checkboxes in the desktop world. Both methods use the drag gesture to activate and deactivate the control; the switch can also be activated by a touch. These elements can be placed anywhere on the screen, but require enough padding around them so they can be easily accessible by a finger. A good rule of thumb is to provide 20 to 40 pixels of padding of empty space around these elements.

The Tab Bar

iOS 6

iOS 7

The tab bar is one of iPhone's signature navigation elements for apps. This replaces the traditional tabs found on most web pages with a language of icons and small titles. This element is commonly found in all of the apps that come preinstalled with the iPhone and is available as a standard control. As a result most iPhone users are familiar with this interaction and its behavior. It has several builtin behaviors that make it easy to use and implement.

1. A selected state—Icons are automatically colored blue to show they are active.

2. Consistent placement—The tab is locked to the bottom of the screen on each view regardless of the screen size.

3. Scalability—The tab bar automatically scales itself for retina and non-retina displays. When working on an app to support a retina display icon the designer only needs to scale the image twice as large and add the @2x to the icon name. Add into the project folder and its ready to go!

iPhone Screen Size with Tool Bar.

4. Touch area—Each icon added to the tab bar includes a large active touch area hence the shadowed area around each active blue icon.

5. Maximum width—After five icons the "More" feature allows the user to open a separate view to select other navigation paths.

The tab bar element includes an ability to do some basic color customization; the tab bar background color and active color can be set within the iOS SDK.

The Navigation Bar

iOS 6   iOS 7

Another native app navigation feature, the navigation bar plays a critical role in the experience design of using lists and pages in iOS. This bar is used within the second- and third-level screens to allow for navigation, back buttons, actions buttons, and titles for accessing content. Like the tab bar, the iOS SDK provides basic customization of colors and gradients.

The navigation element always remains consistently placed at the top of the screen, allowing the user to have the experience of finding a way through their content. The basic interaction rules for this element are as follows:

iPhone Screen Size with Nav Bar.

1. Left Area—Button to go back one step or page view.

2. Center Area—Title of the current content or page view.

3. Right Area—Actions button for the current content or page view.

Surprisingly, this iOS app user interface element and interaction has now become popular in mobile web design as well. The design and interaction pattern has been translated for use in the mobile web browser. For the mobile web, the designer will need to create all of the elaborate buttons states (over, on, down) and functionality in order to use it. A more detailed usage of this experience pattern can be seen in Chapter 6 (see the List Pattern).

The Tool Bar

iOS 6   iOS 7

The tool bar acts as a general placeholder for icons, buttons, and text. The goal of this element is to use it to support the current view or page by providing secondary navigation. A perfect example of this is its use within the Safari app. In this example, the tool bar allows for functionality like bookmark, share, and change browser views. Again, this element is always anchored to the bottom of the screen. This element allows for the most customization as it can be changed to not use the gradient background and use an image as its background instead.

iPhone Screen Size with Tab Bar.

Action Menu

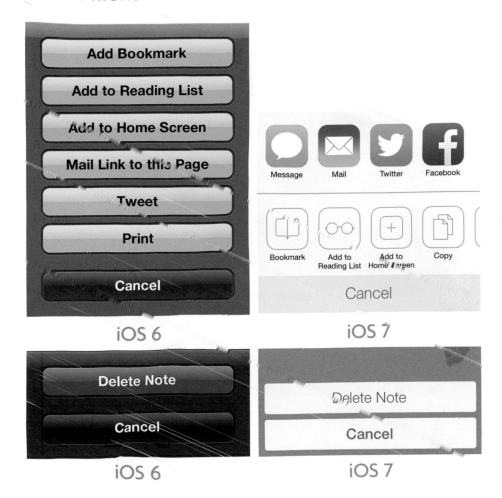

The action menu allows for secondary navigation within an app. The typical scenario is to launch the action menu from a navigation or tool bar. This gives users access to secondary actions or navigation. This element cannot be customized, but does allow you as the designer to add as many different actions as posvsible.

GETTING TO KNOW ANDROID
WELCOME TO THE ANDROID PARTY

I am always asked, "How do you describe the Android?" Taking a moment I respond with this answer.

Android is the phone for everyone. It's the hipster, it's the partygoer, it's the quiet student in the corner, it's the loud cheerleader, it's the drunken uncle, it's the phone for every person. …

This is Android.

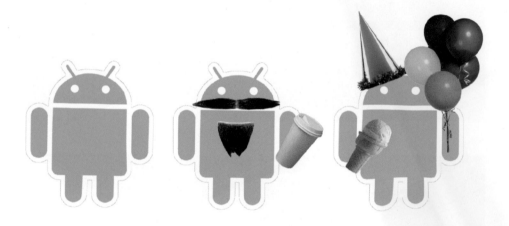

How Is it possible to make one smartphone that can work for everyone? How is it possible that one smartphone can fit everyone's needs, wants, and desires? Simple, you literally make one smartphone for *every* kind of person.

Sample of Different Android Smartphones and Tablets.

I wasn't kidding ... by the time you are reading this I personally will have catalogued over 400 different models of Android phones and tablets. These include every configuration, screen size, form factor, size, carrier, and manufacturer imaginable. In one way or another, an Android phone will appeal to everyone. *This is Android.*

Official Logos for Android Operating Systems. ... No I'm Serious!

To technically answer the first question is much simpler. Android is an open source platform created by Google. Its open source license allows any company or individual to design, build, or even install the operating system on any device. If you are a company building a smartphone why would you create your own operating system when you can license one for next to nothing. That being said, this is one of the reasons Android devices have exploded with all kinds of variations. Every manufacturer has built upon the Android platform to create their flavor of the operating system as well.

a. HTC, b. Google, c. Samsung, d. Motorola.

With every new manufacturer comes a new and exciting flavor or skin of the Android operating system: Samsung has Touchwiz, while HTC has Sense UI. Each includes its own user interface elements, home screen, apps, widgets, and interactions. This makes designing across all flavors an exercise in juggling design and user experience. Are you ready for the challenge?

HARDWARE INTERFACE, ANDROID GESTURES, AND BEHAVIORS

Before moving forward, let's take a step back. Every variation on Android starts from the same base code. The Android platform is based on creating some core interactions and interfaces that all variations share in some way or another.

Hardware to Software Interface

Google   HTC

Samsung   Motorola

Android Hardware Example.

Starting with the Android hardware most devices share some method of hardware buttons. Some are more recognizable than others, but they all try to retain the same functions even across manufacturers and Android flavors. These hardware buttons activate the user interface and can be used for your purpose to design and complement your user experience. Be aware that there is no consistency to the availability of these buttons on Android devices. Some manufacturers will choose to add or remove hardware buttons depending on their own device. Below is a list of what is available:

- Back button—works as a traditional back button in the browser and also works in moving back to another screen in the OS or an app

- **Menu button**—opens the device settings, but works to launch an in-app menu that you can design

- **Search button**—works to open the devices search feature, but it can be used to open a search function with an Android app

- **Home button**—works to return the user back to their home screen

- **Recent apps**—Works to switch from all open and recent apps

When designing an Android app these buttons can be mapped to create an interface for your app. For example, mapping the search button to open a search function within your mobile app experience. This assumes that all of your users' Android devices have a search button; this might not always be the case.

Android Behaviors and Interactions

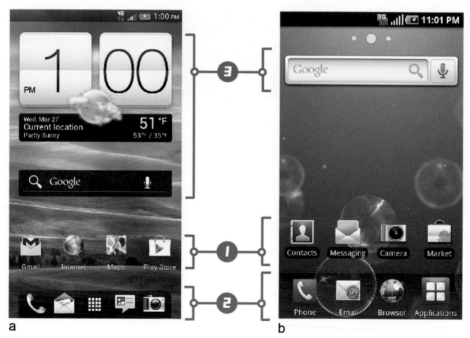

a. HTC Home Screen, b. Samsung Home Screen.

Mobile User Experience | Patterns to Make Sense of it All

The Android interaction concept closely resembles that of the icon-landing page presented in iPhone. Some of the differences in behavior and interaction can be seen on the home screen.

1. Apps—Android also uses the concept of apps, but unlike iOS, these apps create shortcuts on the home screen. The actual app lives in a separate applications screen.

2. Bottom Navigation—Like iOS, the Android home screen adds a tab bar to house the dialer, some apps, and a shortcut to the application screen. This visual styling of this bar is one of the most commonly changed UI elements in the different Android skins. Some tab bars are customizable, while others constrain the number and types of functionality and apps that are displayed.

3. Widgets—A unique UI element specific to Android. A widget allows apps to load data and functionality onto the home screen. Widgets can be designed and built in different sizes. They are included as part of the app and can be loaded onto the home screen after the app is installed.

The Menu

Another element unique to the Android is the menu. The menu allows for additional secondary navigation for an app. As discussed in the hardware interface section, it is commonly opened using the physical menu button found on the phone. This function is being deprecated in Android 4.0 in favor of an action menu, but it still allows for access to the menu for legacy apps.

Touch Gestures

The iPhone created a wave of excitement about touch gestures; the pinch and zoom, the swipe, and the tap became synonymous with smartphones. The first version of the Android OS was released with zoom in and out buttons for the map and browser. Later the pinch in and out were added. Not to say that the iPhone lead the Android to integrate these gestures, but we can say it was an influence. Many of the iOS gestures can now be found within the Android OS; they share the same interactions and behaviors. Like the iOS shake gesture Android also has its own unique touch gesture.

2-Second Touch Gesture.

The 2-s touch is a gesture that most users are not familiar with. Think of it as using the right mouse button on your desktop; it reveals an entire secondary set of tools or navigation. Traditionally it opens a popup menu over the current screen. Surprisingly, this gesture has lasted the test of time and continues to be integrated into the most current Android OS. When developing an app you can apply this gesture to your design elements. Next time you play on an Android device try the 2-s touch.

UNDERSTANDING THE ANDROID KIT OF PARTS

Designing for Android has always been considered the "wild, wild west" of mobile design. Its guidelines have always been more suggestions than requirements. Its design documentation has been close to nonexistent and its user experience guidelines have been a thing of fiction. The library of user elements is limited. It is easier and faster to create your own custom UI elements than to use the UI library from the Android SDK. Unlike iOS, there is no approval needed to get onto Google Play (Android market) and no review process to do so. In some cases, a flavor of Android will even launch without Google Play installed.

With that said, I will give you an introduction to working in and around these obstacles. Once you know the obstacles you can better plan your design decisions.

Screen Sizes

The Android screen size has been a major part of the inconsistent user experience in Android. Every manufacturer, device, and carrier has its own screen size. Take, for example, Motorola which for a time only released smartphones with a 480×854 screen size. Why the extra 54 pixels? … The world may never know. Below are the three most common screen sizes with 480×800 being the most common.

a. 320×480, b. 480×800, c. 540×960.

But wait there's more … Here is a partial list of the screen sizes you may encounter in Android.

- 240×320

- 320×480 (ldpi)

- 480 × 800 (mdpi)

- 480 × 854

- 540 × 960 (hdpi)

- 720 × 1280

- 800 × 1280 (xdpi)

- *And the list keeps growing …*

When building an Android app the SDK handles these multiple screen sizes by splitting each screen size into categories: *small* (ldpi), *medium* (mdpi), *large* (ldpi), and *extra high* (xdpi). If you want your app to work across all of these sizes you are inherently required to create different layouts and image sizes for each design you create. Planning on designing for the mobile web? The flexible nature of responsive web design is perfect for solving the Android screen size dilemma.

User Interface Elements

As discussed before, the Android user interface library has always been limited to a few elements. Within this group there are a handful of user interface elements that are commonly used. To make this more difficult, these elements are stylized differently depending on the Android flavor/skin.

The Keyboard

a. Google, b. Samsung, c. HTC.

    The keyboard is one of biggest moving targets when designing for the Android OS. Every flavor/skin of Android has its own design, size, and functionality associated with it. Samsung includes a swipe keyboard that allows you to drag you finger instead of typing words; HTC includes a keyboard that allows users to choose special characters and numbers with a 2-s touch. The only consistent behavior is that the default keyboard will open when you access a mobile web page or app; it will also be anchored to the bottom of the screen. When designing an app you can control when the keyboard opens and if you can, open a dialer as well.

HTML 5 CODE

Only a few options are available in HTML 5 for Androids' default browser. By setting the input type as you see below you are able open the different keypads, keyboards, and pickers.

- Opens a Telephone Keypad: `<input type="tel"></input>`

- Opens a Number Keyboard: `<input type="number"></input>`

- Open a Date Input: `<input type="date"></input>`

- Open a Time Input: `<input type="time"></input>`

- Open a Date & Time Input: `<input type="datetime"></input>`

- Open a Month Input: `<input type="month"></input>`

Want to see what your Android device supports? These options will vary by Android version and browser. Open this link from your device's browser and select the input fields.

http://www.mobileuxbook.com/keyboard.

Tabs

Android Tab Bar.

Samsung Tab Bar.

The tabs element is unique to building Android apps. Its concept is very simple; allow the user to create a tabbed navigation. It includes some built-in behaviors such as a hover state and an active color for a tab. Unlike iOS, the tab bar can be placed anywhere on the screen. There is only one problem with this element when using it; it inherits colors and stylizing from the Android flavor/skin of the device they are on. For example, if using it on a HTC device a green color will be set to an active tab, when on Samsung a blue color is inherited, and so on. Now imagine designing blue icons and seeing them disappear on Samsung or specifying your text to be white and seeing it disappear on the base Android tab. In this case, designing your own custom control will make your life easier.

Lists

Android List Element.

Samsung List Element.

HTC List Element.

Another app-only element, a list element, best shows the problem of the inherited colors and styling. You can use a list element as a button, a check box, or a switch, you can also add an icon and two lines of text. One of the problems of consistency is seeing the sizes and shapes of the inputs change depending on the flavor of Android. As with the tabs, color also plays a critical role in trying to maintain a consistent UI design; if you make your icons green they will disappear on a HTC device, if you make them blue they will disappear on a Samsung device, etc.

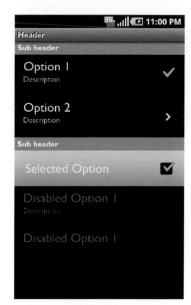

Samsung List Element with Different Input Options.

TURNING OVER A NEW LEAF ...

With the release of Android 4.0 and beyond, Google introduced guidelines for design and user experience. The idea is to reintroduce the Android with a consistent and flexible user interface.

- What has changed?—The menu button is now replaced with an action bar in the UI. The search button on the device is gone.

- What has been added?—An entire library of user interface elements. A library of design patterns. A style guide for colors and typography. Support for tablets.

- What hasn't changed?—There is still no review process for your app, and it is still difficult to make multi-screen experiences across the same app.

Explore Android 4.0's new design and user experience guidelines at: http://developer.android.com/design/index.html.

UNDERSTANDING EVERYONE ELSE

Why did it take me this long to answer the question: "Why only show iPhone and Android first?" This has nothing to do with favorites, which one I owned first, my experience, or even what platform was introduced first or second.

In a nutshell, Android and iPhone have the largest adoption to date.[1] Their basic interaction is easy to understand and similar, both use the same interaction concept of apps and icons, and both are based on the same webkit browser (soon to be changing with the arrival of Google's Blink browser engine). Building a mobile experience across both is easier when you can apply the same design patterns, user interface elements, and user experience design.

That being said, I will do a quick review on the smaller adopters in the mobile space.

Windows Phone 8

Microsoft reinvented its mobile offering with the introduction of the Windows Phone platform. Starting with their Version 7, they took a different approach to the interaction concept of apps and icons. Their approach focuses on a more visually designed experience with an emphasis on text and images. The layout of an experience is designed to spill into other views, creating a fluid and organic layout when travelling between pages. In some cases titles, images, and page content overlap each other on the screen. Designing for this experience requires designing patterns that overlap

---

[1] Chris Burns, "Nielsen 2012 recap puts Android and iOS on top," December, 21st 2012, http://www.slashgear.com/nielsen-2012-recap-puts-android-and-ios-on-top-21261981/.

multiple screens compared to separate views or pages as in the Android or iPhone. One example is an app that might scroll horizontally compared to scrolling vertically within the Android or iPhone.

Taking a cue from Apple and the experience from the Android, Microsoft created standards for hardware and user experience guidelines to complement the creation of apps for their platform. Yet, with all of this support, market adoption has been slow due to limited devices released and software support for the platform.

Nokia Symbian and Beyond

One of the larger players internationally, Nokia's Symbian platform has dominated feature phones around the world. Yet, the Symbian platform for smartphones has not had the largest adoption compared to iOS and Android.

Nokia Lumia 920 Phone Running Windows Phone 8.

Looks familiar? In 2011, Nokia announced a partnership that would install the Windows Phone platform on new Nokia smartphones. By 2013, Nokia officially killed Symbian in favor of using Windows.[2] With this support the Windows phone OS could potentially become a third running mate in the mobile race. I'll keep my fingers crossed for them.

---

[2] Matt Warman, "Nokia ends Symbian era," *The Telegraph*, January 2013,
http://www.telegraph.co.uk/technology/nokia/9824179/Nokia-ends-Symbian-era.html.

Blackberry

I would call this the story of "riches to rags." Once the dominant player in business smartphones, Blackberry had led the largest adoption of early smartphone use. Its reach was vast, but its focus on user experience was nonexistent, leading to a fragmentation of different devices, form factors, and operating systems, making it close to impossible for any designer or developer to keep up with the changes. Features that existed in one OS version disappeared in another; touch gestures that were created for one device would only be useful for that singular device model. Now Blackberry devices struggle to keep a small market share. In a world now dominated by the iPhone and Android, the story of Blackberry is about trying to do too much across too many devices at once (Sounds familiar … Android?).

GETTING TO KNOW THE TABLET

Why did I not include tablets in this conversation? Simply because, tablets are NOT smartphones and they are NOT desktops, they are their own unique experience. A tablet lies between the bite-sized experience of the mobile and the full-sized desktop counterparts. These live in a gray area that is in the process of being defined. To design for a tablet is to ask the question: what will you NOT do on mobile and what will you NOT do on the desktop. This is the tablet experience.

With a much larger screen size (1024×768) and a much larger weight (anywhere from 1 to 2 pounds) compared to the mobile device, the inherent interaction and behavior of using a tablet is different. Users commonly use two hands to engage the tablet; one hand to hold and the other to touch. Most tablets are WiFi only, but the trend to make them network-capable devices is starting to advance. Another movement is that of making a smaller tablet with a 7 to 8-inch screen.

As user experience designers, our goal is to design content and interactions on one or two mobile screens. With the larger screen real estate and engagement of two hands, we are able to create and drive more functionality on only one screen. Tablets are normally rotated more than mobile devices. As a result, the user can view pages with the 768 pixel width or a 1024 pixel width with just one turn.

I will review the two most common tablet operating systems; you might have guessed by now these will be based on the iOS and Android.

iPad (iOS)

iOS Tablets (iPad & iPad Mini).

The iPad was first introduced as an extension of the iPhone operating system. It shared much of the same gestures and UI elements that were well known in the iPhone world. Users did not have to learn a whole new set of gestures or interfaces as the iPhone had already taught them how to use these. To say the least, it was like using a big iPhone. Since its release the iPad has become its own entity. Now with more features and functionality (front facing camera, retina screen), it uses some iPad-only multi touch gestures and UI elements. For example, three fingers allow you to switch from one app to another; recently a split keyboard was added to allow iPad users to type with both hands. Apps can be designed and built using the universal format to create an app that opens on

iPad Split Keyboard.

both the iPhone and iPad. The developer and designer will need to build a separate view for the iPhone and iPad, but they will be encapsulated within the same working file.

Android

A more interesting story in the world of tablets, Android first released a tablet-only version of their OS, Android 3.0 (Honeycomb) also known as the forgotten operating system. Designed under the concept of tablets being unique compared to the behaviors and interactions of a smartphone; the OS was a failure. The OS was so unique that its UI elements and interactions would only work on apps and experiences that would be specifically developed for 3.0. One of the positive results of this endeavor was the reinvention of Android 4.0 to create a consistent user experience across all Android devices.

Android 4.0 Smartphones and Tablets.

Both tablets and smartphones now share the same UI elements and user experience behaviors. Though it's a preliminary attempt to unify these experiences, time will tell if this helps to lessen the effect of multiple screen sizes and form factors in the Android tablet world. When building a 4.0 app the Android designer and developer still need to build using the traditional layouts (xdpi for tablets). The user experience behaviors for how to handle larger multi-screen layouts is now part of the new Android 4.0 guidelines and documentation; a much-needed improvement for the Android.

PUTTING IT ALL TOGETHER

Now that we know the devices and differences in mobile platforms, we can start with an informed view on the mobile. This may seem like a difficult challenge, but I think otherwise; the best part of a mobile is that we have some set boundaries as a starting point. The challenge is being able to design with a unique frame, a unique set of UI elements, and a unique set of gestures. These are all constraints that are easy to overcome with just a little bit of extra work. Our job as experience designers is to keep our pulse on this ever-changing ecosystem. I see this as a positive, by keeping up with this changing mobile landscape we can actively contribute to the innovation of the mobile from an informed perspective.

NEXT STEPS: BUILDING THE NARRATIVE

The start of any good narrative is to introduce the cast of characters, places, and events. Before we can start to form our mobile experience, we need to learn everything about the context that these experiences will live in; what different devices are like, how gestures work, and what makes their UI elements unique. Think of the devices as our cast of characters and the operating systems as stage sets. Next, we need to learn how to write a compelling story and how to visualize it. For our mobile user experience this will be our wireframes. The dialogue will be our mobile patterns and the performance will be our use of prototyping. All of this combined will create the narrative we present to our mobile users. Some narratives started with "Call me Ishmael" others with "It was a dark and stormy night," but ours starts with …

"Be on the device."

CHAPTER 5

How Mobile Wireframing Works

INTRODUCTION

The goal of the wireframe is to plan and set in motion the format of the mobile experience; here is where you lay out the groundwork, intent, and interaction of your experience. Before starting the wireframe process, you will need to define these criteria for your experience:

- Screen sizes

- Platforms

- Interactions

- App or Mobile website

- Smartphone or Tablet

We will be using the wireframe not just to define sizes, content, or interactions, but also to tell a "narrative." How will your user enter the experience? What will they do there? How do different platforms affect the experience? How does it work on a tablet versus a smartphone?

By building a mobile wireframe, we will be able to document the narrative and intent of our experience. By doing this correctly, we will make it easier for a visual designer to create the UI design and the mobile developer to build it.

As we move forward, we will look at how selecting our mobile criteria is important to your first wireframe layout. Once we lay out our first wireframe in this chapter, we will be building it up over several layers of information. Think of it as a work in progress …

… Now let's start wireframing!

CREATE CONTEXT

Unlike wireframing for the desktop, mobile development requires us to set some context to the narrative we are building. Think of this as a stage set for a play. If you are designing for Android, use an Android frame, if designing for iOS, use an iPhone frame, and if you are designing for the mobile web, then use a more generic device frame. As our experience will always be viewed within the frame of a device, this helps visualize and contextualize your experience to the device. For example, if you are designing an Android app to use the hardware buttons you will need to make reference to these when you wireframe your experience. This is much easier when you include the frame within your wireframe.

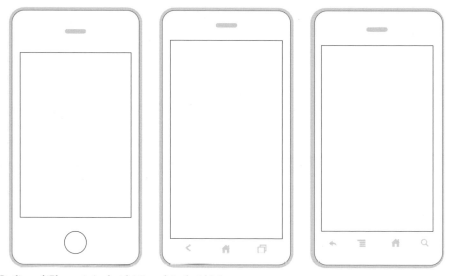

Outline of iPhone 4, Android 4.0, and Android 2.1.

Image of iPhone 4, Nexus One, and Windows Phone.

Visually, the design of the frame is up to you. You can either use something more literal, like an image of the device, or more stylized, like an illustration of the frame. In my experience, I have used both. Using the image of the device creates a strong visual impact when you present the wireframe. ... It says, "this mobile experience is for an iPhone." Using a stylized illustration for the device is easier to work with when you have a large wireframe deck; this helps reduce the file size. When creating an experience for the mobile web it is much easier to use a more generic illustration, an image of a device everyone knows as it clears the way for the client to imagine the design.

TOOL TIP

Want to know how to make a device frame for yourself?

Here are three methods.

Method #1: Go online and find an image of an iPhone or Android device. Open the image in a photo editing software and cut out the screen size. Scale accordingly, save, and presto! (Make sure you keep it at 100 dpi or higher if you want to print out your wireframes. If you are planning to only share by email, then keep it under 100 dpi.)

Method #2: Go to the SDK websites for Android or Windows. You can download files for device skins (images of phones) there.

Method #3: Using a downloaded image, trace it in vector illustration software. Save it as an EPS file format; once completed you can scale and shrink this vector artwork.

Screen Size

The next step to building your wireframe layout is to select a screen size. If you are looking to cover all of your bases, select a common resolution like 480 × 800 pixels. If you are looking to work for iOS select a screen resolution like 640 × 960 pixels. Make sure the selections you choose will correspond to your business decisions on what platforms or operating systems you want to support.

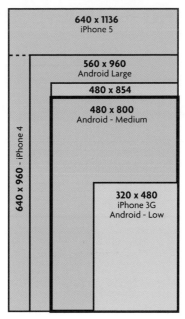

640 x 1136
iPhone 5

560 x 960
Android Large

480 x 854

480 x 800
Android - Medium

320 x 480
iPhone 3G
Android - Low

640 x 960 - iPhone 4

Mobile Screen Size and Proportion Matrix.

Selecting a screen size is as much about size as it is about proportion. The proportion/size matrix shows a quick snapshot of how sizes and proportions change with size. If you are looking to cover a wide range, you will need to think about an elastic design versus designing to only one target. In reality, even if you are designing to an app or mobile website you will still have to support some level of multiple screen sizes; this is inevitable. As we build our wireframes we will look at how to respond to various sizes and how to represent them.

Once you have selected your target screen size, make sure that your device frame works with your selection. If you have chosen a screen size of 640 × 1136 pixels (iPhone 5) make sure your device image or outline matches.

REPRESENTING PAGE AND SCREEN FLOWS

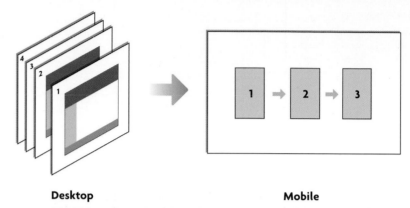

Desktop   **Mobile**

Desktop Wireframes versus Mobile Wireframes.

One of the unique characteristics of mobile wireframing is the small size of screen real estate. As a result, the language of showing actions is much harder if you separate screens as separate pages. One of the techniques to combine actions into a flow is to connect three to four screens together. In practice we would lay out three screen frames on each page. Using arrows, we want to show how interconnected paths get displayed as mobile screen flows. Think of a set of mobile wireframes more as user flows than pages. Each user flow will then be separated on an individual page. As a result, each page will tell a story of a user action or function.

CREATING YOUR FIRST LAYOUT

Now that you have chosen your screen size, device skin, and learned about how a mobile set of wireframes will flow, let's start assembling a page! Here is a list of things you need to know as you create your first layout:

1. Use a large page size. I would recommended a 11" × 17" at a minimum.

2. Use your device skin on the first panel. It is there to give the reader context, not to clutter the page.

3. Name the flow. For example, add to cart, add a friend, etc.

4. Name the app. Be clear if this is a native app or mobile web app.

5. Create your first page as a template. You will be reusing this page over and over again, so make sure you can reference it.

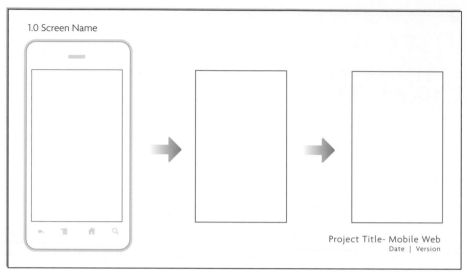

Example 1.

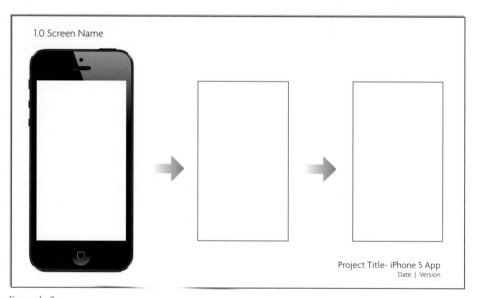

Example 2.

Mobile User Experience | Patterns to Make Sense of it All

TOOL TIP

What software should I build these mobile wireframe templates in?

Building your wireframe templates can be done in any software. I have used a collection of different software depending on the project. I have used InDesign, OmniGraffle, Visio, Keynote, and even PowerPoint. Whatever you choose, make sure you account for two criteria that will make your life easier:

1. Allow for multiple page layouts.

2. Export to a format you can email and print (like PDF).

REPRESENTING INPUTS

One of the most important differences between designing for the desktop versus mobile is the use of the device input as key design elements. On the desktop we don't worry about the users' keyboard or even what a dropdown will look like, these are all standardized. In wireframing for mobile, we have to take into account the device and platform keyboard type and size, and any standard controls of UI library elements that you may design with. In our case, we need to use the actual elements for our design. For example, if you are working on an iOS app, then specify the actual iOS keyboard; our goal is to separate any prebuilt patterns and UI elements from those that you are designing. This makes your presentation read more realistically and announces to the designers and developers what is already handled by the SDK or device, compared to what they need to design and build from scratch.

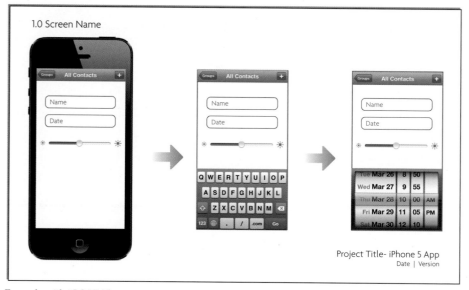

Example with iOS UI Elements.

TOOL TIP

Want to know where to download wireframing stencils?

Don't reinvent the wheel. There are resources online to allow you to download UI stencils for different formats and platforms. Here are a few places to download them:

General Resources

www.mobileuxbook.com/wireframing

www.graffletopia.com (for OmniGraffle only)

Android

http://developer.android.com/design/downloads/index.html

iOS

http://www.teehanlax.com/blog/ios-6-gui-psd-iphone-5/

http://www.teehanlax.com/blog/ipad-gui-psd-retina-display/

Windows Phone

http://blogs.windows.com/windows_phone/b/wpdev/archive/2010/07/27/windows-phone-7-design-resources-ui-guide-and-design-templates.aspx

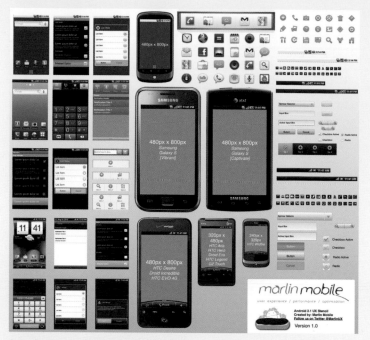

Sample Omnigraffle Android UI Stencil.

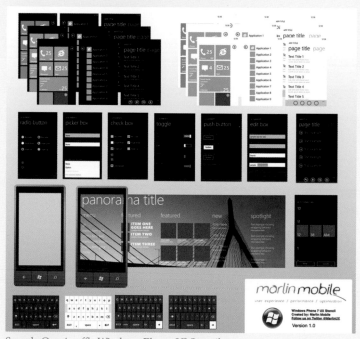

Sample Omnigraffle Windows Phone UI Stencil.

REPRESENTING GESTURES

As you lay out your mobile user experience you will find cases where you design an interaction that will require you to represent a touch gesture. In these cases you will need to add indicators of these gestures to your set of wireframes. In my example, I use illustrations of the gestures with and without fingers. They are interchangeable, but each has its own advantages. Showing gestures with a finger makes it very apparent to the reader that the gesture is to be controlled in a very specific manner (i.e., pinch in or out). It also highlights a prominent action that will take place in the user experience. Using illustrations of gesture without fingers makes it easier when working on an experience that will require the user to use several gestures (i.e., in a game); in this instance using a simple gesture will make it less distracting to the reader. You can build your own library of touch gestures, just make sure that the colors and visual style you use remain consistent.

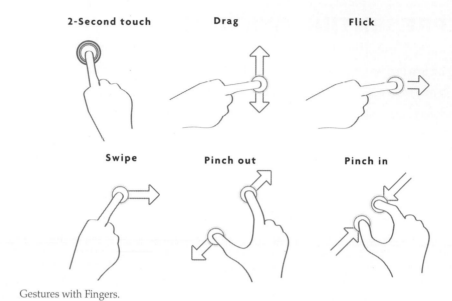

2-Second touch   **Drag**   **Flick**

Swipe   **Pinch out**   **Pinch in**

Gestures with Fingers.

2-second touch   **Drag**   **Flick**

Swipe   **Pinch out**   **Pinch in**

Gestures without Fingers.

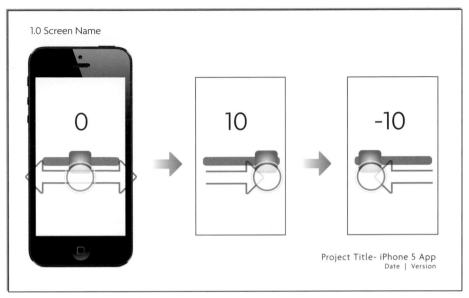

Gesture in Wireframe Example.

ANNOTATE, ANNOTATE, ANNOTATE

One of the most important pieces of a wireframe deck is annotation. Remember you are not only laying out your design, but also providing instructions. When designing mobile experiences it is critical to denote every action and interaction you want to capture. Here are three common examples of what you will need to capture:

1. Annotate UI actions—If you want the keyboard to open when the user first opens the app, make sure to create an annotation under the first screen.

2. Annotate Layout—If you want your background to stretch and be elastic, or have an element anchor to the top or bottom of a screen.

3. Annotate Gestures—If your experience requires multiple gestures, be clear on what UI elements are activated by the gestures.

Don't worry if you need to add an extra page for your annotations. Our goal is to capture as much relevant information about your screens as you can add. I usually wait to do this at the end of a design session. Once I know I have fixed my UI elements and page interactions then I go back and document them.

WIREFRAMING OTHER CASES

When representing a gesture and adding annotations are not enough, you will need to create a new layout that does not fit snugly within the three to four screen flows. In these cases, you will need to adjust your layout to represent this new page format. I will present three of the most common cases; representing the use of the accelerometer, working with multiple screen resolutions within the same app, and responsive web design.

REPRESENTING MOTION (ACCELEROMETER)

One of my favorite aspects of working with mobile devices is being able to design using the accelerometer. This allows you as the user experience designer to have the devices motion or rotation start an interaction. As you can predict, this will likely need its own layout. Our goal here is to be able to show how motion via the device or by the users own gesture will control the start or end of an interaction. In these cases we will need to use either illustrations of the motion or arrows to denote what is actually being moved. Use a layout to describe the before and after state if you are using a rotation or the active state if you are interacting with the device (i.e., tipping the device left or right).

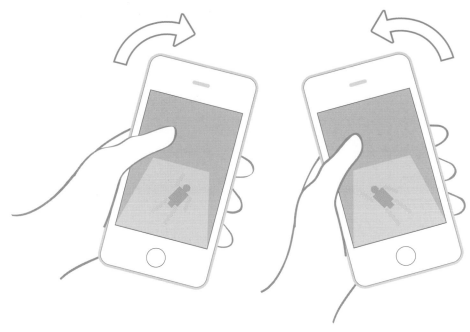

Example Showing Accelerometer.

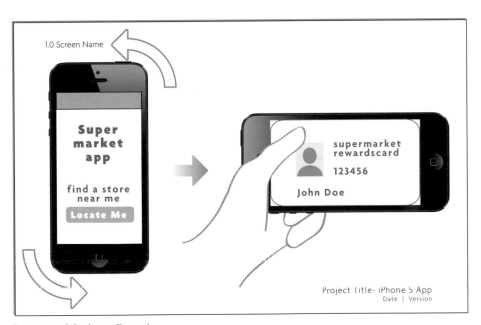

Rotation in Wireframe Example.

REPRESENTING MULTIPLE DEVICES (MOBILE APP)

When designing an app you will find yourself having to fit your experience into several screen sizes. This case is more common than you think. One of the typical cases is adjusting the screen for iPhone 5 when working on an iOS app. As the screen expands longer than the typical 640 × 960 resolution, you will need to show how your experience will be adjusted for the difference. Another iOS example is designing a universal app that is both iPhone and iPad compatible; in this case you will need to show how the design relates to both formats. When working with Android, you will need to design your experience for the different screen sizes (small, medium, large, etc.). In these cases you will need to use a combination of wireframe examples and annotations.

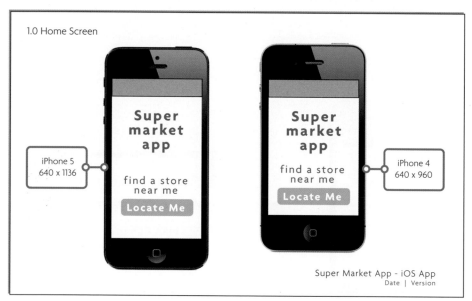

Example Wireframe for Multiple App Screen Size/Devices.

REPRESENTING RESPONSIVE DESIGN (MOBILE WEB)

Designing for responsive web design has several challenges, one of which is representing how different break points (compatible screen widths) get illustrated in your mobile web wireframes. In this case, you will first need to define all the screen widths you are planning to support. Will you end up designing for a smartphone, tablet, and desktop browser? If so, plan on what screens will be responsive and design around them. In a typical case you can show three sample screens on one page; the idea is to show the relationships between each different width.

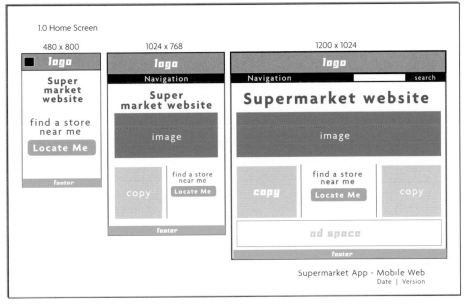

Sample Responsive Design Wireframe Sheet.

Are you planning on changing the user experience of each break point? If you answer yes, I have some bad news. Plan on creating separate decks for each break point/ screen width. Because each experience will be dependent on the unique screen flows, you will need to represent these as separate wireframe decks.

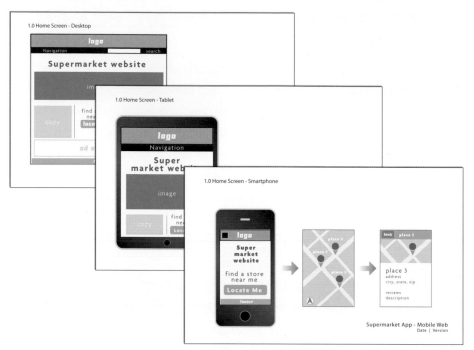

Sample Full Responsive Design Wireframe Deck.

Mobile User Experience | Patterns to Make Sense of it All

CHAPTER 6
Mobile UX Patterns

WHY PATTERNS?

Many years ago, I had an opportunity to sail with Ted, a friend and retired IBM software engineer. When I asked him what he wanted me to do on the boat, he replied, "I want you to sail the spinnaker." To my knowledge, this one of the most complex and difficult pieces to master (in fact, I had no idea what I was doing when I started); when I asked him why he wanted me to sail the spinnaker, he replied with this: "You're a UX designer. Trust me, you'll figure it out. Just try to match the pattern on the sail as it moves in the wind."

What Ted had been referring to was the movement of the top quarter of the sail, a piece no more than 12 inches in length and some 15 feet away. As the boat accelerated and decelerated, the shape of this small portion in the sail would change in the wind. By matching a certain pattern of movement to the sail, I could make the boat go faster or slow it down. The more time I spent sailing, the more I started to experiment with the patterns that the wind made in the shape of the sail. What happened to this piece of the sail in high wind versus low wind, when the boat was at a different angle and even in different wind speeds? Ted had been right: this was a perfect job for a UX designer.

When I first began thinking about how best to describe mobile user experience, I struggled with what to call the library of experiences that I had collected. Are they archetypes? Are they interaction models? Or, are they processes? I remembered my experience of learning to sail with the spinnaker. As I thought more about this, the metaphor of sail patterns made perfect sense; you have a visual pattern or behavior that you are trying to replicate predictably and a series of controls to make these patterns appear. As in sailing, to best know the optimal experience you need to start with a defined set of patterns that are already successful. Once you have a set of familiar patterns working repeatedly, you can begin to change, tweak, and optimize the experience of each; hence the name of this library I have collected: Mobile UX Patterns. As I learned through sailing, with UX design it is important to first decide your intent, use the right pattern, and once it is successful, *then* begin to change, tweak, and optimize to your exact experience.

The goal of this chapter is to present a series of visual patterns and their use cases. I will focus on end-to-end use cases, where we see user interaction, design elements, user experience flows, and information on user intent. I will be presenting them in forms of illustration, wireframes, and sketches. The intent is to show how these concepts work and how best to represent them as visual experience patterns. Some of my patterns will be driven by user interface, some driven by user action, and others driven by the hardware; regardless, these patterns allow mobile UX designers to create an experience for the users via different methods and techniques to meet the user's goals. I have collected these patterns into a library; each UX pattern is different from the other, but they can also be combined to reinforce one another. As you continue through the chapter, you will see how some patterns may complement or affect each other. Feel free to mix and match patterns. As you do, remember the metaphor of the spinnaker sail: decide your intent, use the right pattern, and once it is successful, *then* begin to change, tweak, and optimize to your exact experience.

As you go through the patterns, I provide examples to some of the wireframes and visual elements included in each. I add annotations to patterns and elements that you can use in building your own experiences.

PATTERN 1: THE LAUNCHER

This pattern—let's call it the launcher or springboard—is based on a simple concept: give the user a screen where they can access various applications, functionality, and actions. Most users of Android and iPhone devices are accustomed to the interaction model of a home screen with apps that they can launch when touched. Our goal with this pattern is to mimic this learned behavior and apply it to our design so the user can navigate your mobile experience.

THE BASIC LAYOUT

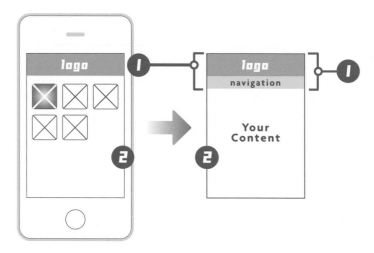

This pattern is made from two main components.

1. The header

2. The work area

The Header

The header is used to display two basic functions. The first is the branding area for your app or website and the second is the navigation area. Our goal with this area is to allow consistent navigation and branding across the experience. As with designing for the web, consistency allows your user to have a level of way-finding in an experience; the mobile is no different. As we dig deeper into our content pages, we can use this area to include navigation elements, for example, a back button in iOS or navigation tabs in Android, all of which are ways to navigate back to your main content areas.

The Work Area

The main goal of the work area is to allow you a flexible place to house the content of your experience. As the navigation and branding are combined in the header, the work area gives us a clean slate to focus on our content. For business apps we can use the work area to house templates for functionality (e.g., bank for checking my account), for media we can use the area to house templates for content, and for games we can use this as a springboard for different levels and mini games.

Visually, the work area requires us to provide flexibility to the design. Our experience needs to be fluid in not only the width, but also the height. The experience should be able to scale to multiple screens proportionally to handle the jump from 320×480 to 480×800 pixel screen sizes. It should also be able to change its height when the experience requires or the orientation changes. This may seem an unusual use case, but this is now very common, for example, when creating an experience to fit into an iPhone 4 versus an iPhone 5, where only the height of the screen has been enlarged. This is also the case with some Motorola Android devices that now have a uniquely larger screen height (854-pixel vs. the more common 800-pixel height). When using this pattern for the mobile web, using responsive web design can solve these use cases; flexible page widths can be used to scale across the multiple screen widths and heights.

Creating Connections

The true power of the interaction of this pattern is the creation of connections for the mobile user. Here are some examples of what I would call a true mobile "integration."

- ■ Example 1: Use a launch icon to start the dialer.

- ■ Example 2: Use a launch icon to start the camera.

- ■ Example 3: Use a launch icon to open a web page from a mobile app (this would launch a window of the WebUI view).

- ■ Example 4: Use a launch icon to download an app from the mobile web.

- ■ Example 5: Use a launch icon to connect one app to another app. In iOS we can set logic in place to allow apps to talk to each other; this is referred to as "in-app communication." If the app is present on the device, the app will open; if not, the app connects the user to the iTunes store. On return, the new app will connect from the launch icon.

ADDING EXTRAS

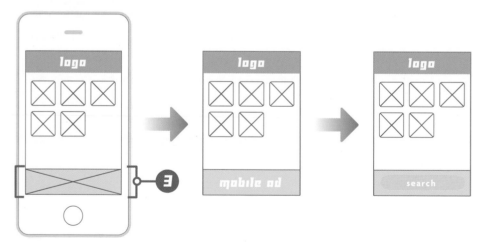

Extra Area.

As we lay out the launcher concept with more features we can focus on adding additional content to the display area. We can look at being able to engage our users with features and functionality to enhance their user experience and meet our business goals. The goal is to create a container that gives us a fixed size to add more depth to the work area. I call this container the extra area (see label #3).

Ads

One of the most common cases is being able to add advertising space to the work area. This is becoming one of the most typical methods to attract users to other apps and websites. In the app world this is called In-App Advertising; users can access other apps for rewards, or be able to advertise your own apps to your consumer base. The goal of using this fixed area is to define the placements of the ads without distracting from the design intent or interfering with the user's goals. If we have created a flexible work area, we can tie the ad to the bottom of the screen and have it move and scale with different screen sizes; this makes our design fully responsive.

Search and Features

Our work area can also be used to allow (or encourage) the users to explore deeper into our experience. We can use our extra area on the home screen to promote content, add a search bar, and display deep links or content directly from the first screen. In this case, I would explore moving the extra area right below the header. This creates a hierarchy of areas for the users to explore as their eyes move down the screen. We can use this area to make the launcher feel more dynamic by periodically changing the content if we wish; making the launch feel fresh when a user revisits the app or the mobile website.

THE ADVANCE LAUNCHER

One of the strongest parts of the launcher concept is the intent to keep the user engaged with different parts of your mobile experience. By adding features or ads in the extra area we can push content to our users. But, what if we want to really enhance their experience and tailor it to their visit? Our next step is to begin to see how dynamic we can make this pattern.

Sliding Panels

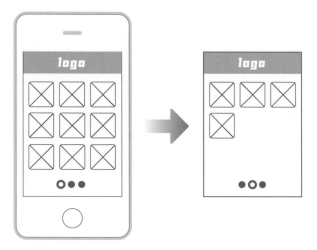

Adding additional panels to the screen can enhance the work area and launcher, simulating the home screens of most smartphones. For apps, we can use the gesture of sliding from left to right and back to access these panels/views with more icons. For the mobile web, we can scroll down the screen to expose a larger number of icons or add swipe gesture with the addition of an external JavaScript library. Taking this further, we can create pagination to simulate the app experience as well. This allows our users to be exposed to more content without leaving the first page of the experience.

Logging In and Out

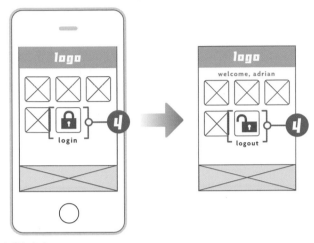

Login Function in Work Area.

Our second advanced feature is to create log in and log out functionality. We can replace one of the icons with a login button. Once the user logs in, we can replace the extra area with unique content or personalization for our users. We can then replace the login button with a logout icon as well. Our goal is to make a seamless experience by having our user add their credentials with little effort. With both apps and mobile websites, we can locally store users' credentials to the device allowing the users to return to the experience without having to log back in. The more we can personalize the launcher, the more frequently they will use our experience.

Personalizing and Customizing the Launcher

Continuing with our personalization trend, we can introduce the concept of having the user customize their experience. Imagine being able to have the user change the look and feel of the experience, or even the amount of icons. Here is list of elements that might be used to give them a more engaging launcher screen.

- Example 1: Give the user the ability to change the color scheme of the experience.
- Example 2: Give the user the ability to hide or show icons on the launcher.

- Example 3: Give the user the ability to change the background by launching the camera or cameral roll.

- Example 4: Give the user the ability to add or remove the panels.

Taking this further, we can experiment with using game mechanics to have users gain points or features by experimenting with customization of the launcher. By making it a game, we can have our user play, try, and customize the launcher. The more we can engage the user, the more we can get them hooked into using and returning back to our mobile experience!

PATTERN 2: THE TRAY

This pattern sometimes referred to as the tray, drawer, or off canvas layout allows the designer to add more space to the experience without the user losing the screen or view that they're on. The main feature is based around having the users use a gesture to either slide or touch to open this tray area.

THE BASIC LAYOUT

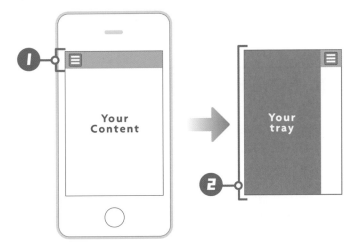

This pattern is made from two main components.

1. The Active Button

2. The Tray Area

The Active Button

This element should always be a visible UI component. Either a button or tab, the intent is to make it clear to the user that this element is clickable and active.

Active Button Examples.

Here are some rules of thumb:

- Make the shape unique compared to your other UI elements.

- Make the design look active (a different color, shading, or texture).

- Make the button look like it is part of the navigation.

- Make sure it's large enough for a finger.

- Make the button look different when it is touched (a different color, shading, or texture).

The Tray Area

The tray is an extension of the experience; use this area to supplement and complement your experience. The design intent is to have the user quickly interact with the active button to enter this tray area. Once there, your goal is to offer them information and functionality to enhance the screen or view they are on (i.e., navigation, tools, or profile information). Make sure the tray area encompasses 60 to 70% of the screen width, leaving some of the foreground screen exposed. The idea is to provide a way-finding approach to making sure the users know what screen they were on and can easily return back to it.

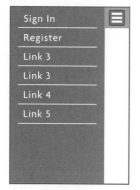 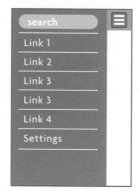 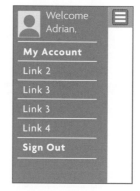

Examples of Different Tray Options.

The Advanced Tray

The tray can be used for showing more advanced features; for example, a tray that is a scrollable view. Some navigation or filters might not fit within the view. Allowing the user to scroll will give them access to a larger tray area.

A second feature is to include other UI elements inside the tray. For example, a log-out button, UI radio buttons or check boxes to turn a feature on or off, or a search field are some options that can be added to enhance the tray experience.

Another option involves launching the tray from a different position on the screen. This gives you the ability to use the tray for loading in content, icons, banner ads, or notifications instead of navigation. This also gives you a wider use of the screen area.

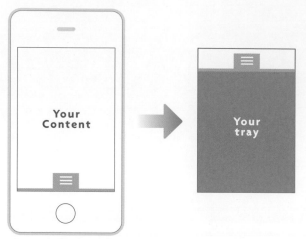

Example of a Tray Opening from the Bottom.

Good Gestures

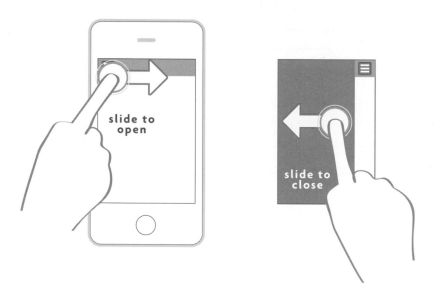

slide to open

slide to close

Along with a clear and visible active button, the use of clear touch gestures to access the pattern should also be incorporated. Your active button should be accessible with the use of a swipe and touch. Use the left gesture to open and the right touch gesture to close the tray. If you follow these rules of thumb for the active button, your button should be large enough for a user to easily swipe the screen.

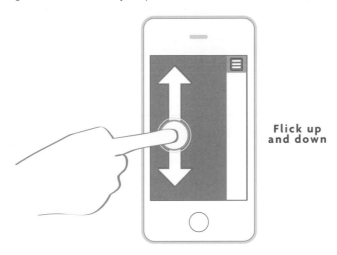

Flick up and down

When using the scrollable feature in the advance tray, the users should be able to swipe up and down to scroll the screen along with using the left and right to open and close.

AVOIDING MISTAKES

A good pattern should provide you the flexibility and support for a positive mobile user experience, but sometimes there are clear use cases to avoid when implementing this pattern.

BE CLEAN AND LEAN

Avoid making the tray area too complicated. This is not another content area for you to add articles or legal text. Make sure the tray is easily readable and simple to engage. If you use navigation or UI elements make sure they are large enough for a user to touch.

KEEP NAMES SHORT

Make sure the navigation and buttons you include are short and direct. If your text wraps to a second line, you might have a problem.

DON'T SCROLL TOO MUCH

If you decide to scroll the tray, keep it to one or two screens, the smaller the better. The user should not feel like they need to deep-dive for content within the tray.

DON'T ADD OTHER GESTURES

Keep your gestures to swipes to open and close and to scroll. Adding in a pinch or a shake won't win you anything with your users, and will confuse them if the application is out of an anticipated context. This is a supplement to the screen, not its own environment, so don't treat it like its own app inside the tray.

DON'T OVER DO A GOOD THING

Just because you can create a tray, don't overdo this interaction; consistency is your friend. By adding more and more trays and active buttons you may confuse the user to the point of not knowing what content you want them to access and why. When in doubt, think back to point #1: Be clean and lean.

PATTERN 3: THE LIST

This pattern is one of the most commonly used in mobile UX. The design is based on providing basic navigation to second- and third-level pages, yet still providing a method to quickly return back to a main screen. This gives a clear indication to the user that they are travelling deeper into a list of pages and viewing page details.

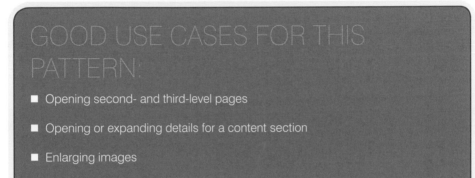

GOOD USE CASES FOR THIS PATTERN:

- Opening second- and third-level pages
- Opening or expanding details for a content section
- Enlarging images
- Opening settings screens or options

THE BASIC LAYOUT

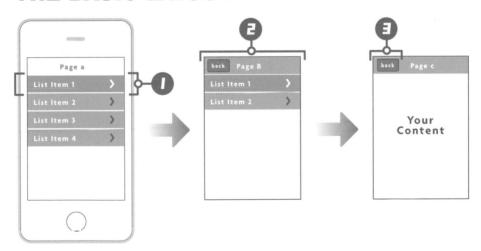

This pattern is made from three main components.

1. The List Item

2. The Navigation Bar

3. The Navigation Buttons

The List Item

The list item should be a user interface button. It can be marked with an arrow, hyperlink, and it can also have an over state to show that the button can be touched to activate. In typical cases we would use these to show detailed descriptions, larger-sized images or second-level pages. As a designer, your goal is to keep these buttons looking clear to the user so that they can be accessible and still maintain a consistent design across the experience (i.e., every list button should have the same look and feel or use of opening a deeper page).

The Navigation Bar

The navigation bar design should be consistent across all the pages. This will provide the "way-finding" for the user as they explore the experience. This bar should always remain consistent when diving deeper into pages and can also provide the current page title as reference. Users will identify with this element to anchor where in the experience they have moved into and how they will return. The navigation bar will also house the navigation buttons to travel back to previous pages.

The Navigation Buttons

The navigation buttons are used to return the user one step backward in the experience. A typical case is to have a "Back" button in the top left hand of the navigation bar. As the user travels deeper and deeper into the pages the "Back" button should always return the user to the previous page.

Mobile User Experience | Patterns to Make Sense of it All

THE ADVANCED LIST

The list pattern can open up several possibilities in how we can visualize content and product pages. In most cases the simple use of the pattern will do; in more advanced cases, look at these two examples of how we can further enhance the pattern.

Making the Last Page Actionable

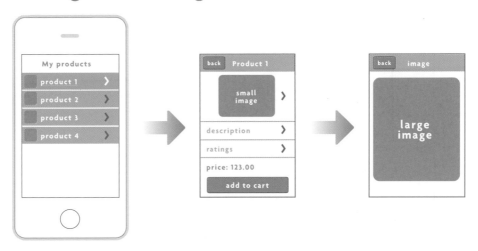

As I have stated before, good mobile user experience should always create an intentional flow. There should not be any dead ends to the users' exploration. To ensure that the list pattern adheres to this goal, we need to make the last page actionable.

We can use two methods:

1. *Provide an action button in the top navigation bar.* In most cases for content rich pages we can use a share function to access social media or email. We can also allow users to save this page for their bookmarks or favorites.

2. *Provide an action button below the page content.* In the case of retail or e-commerce content we can add an "Add to Cart" or "Buy" button to the end of this page.

Using a Dynamic List

A step further to a static list of pages or links is using this pattern to manage a dynamic list. A perfect example of this is using the list pattern to display search results. In this case, a search bar can be used to allow the user to enter a set of keywords and as search results emerge they can be displayed below the search box. By creating these pages or lists as active links, the user can touch each result. As the user previews the results, the navigation bar can have a button that returns them back to the search list instead of a "Back" button. In this case, I would use a "Search Results" button to lead the user back to the list or results. This will allow the user easy access to preview and return back to the search results. The search result details can also be made actionable by giving the user the ability to save the result, share the result, or view more detail.

AVOIDING MISTAKES

DON'T OVER DO A GOOD THING

Try using this pattern to dig into two or three levels deep. If you find yourself digging too deep, you might have to rethink your use of this pattern or the number of levels of pages you have. At over three levels, you will find yourself confusing the position of what page the users are in. Remember that this pattern is about providing way-finding through your content.

DON'T SKIP A STEP

page a   page b   page c

The key in using this pattern is to always keep consistency moving forward and backwards through pages. If you are on **Page A** and enter a **Page B** the "Back" button should always take you to **Page A**. If you are on **Page B** and enter **Page C**, the "Back" button should return you to **Page B**. A bad example is to be on **Page C** and have the navigation bar return you to **Page A**.

NO DEAD ENDS

Just as I stated above, always make your last pages actionable. The worst thing you can do to your users is to make them dig through several pages and have them lose their place. They will leave frustrated and not take the time to return.

PATTERN 4: THE ROTATE

This pattern is based on using the device's built-in accelerometer to detect when the phone or tablet is being rotated from one orientation to another. This pattern is not just about resizing the screen's width and height, but to provide a new experience based on how the user interfaces with the orientation.

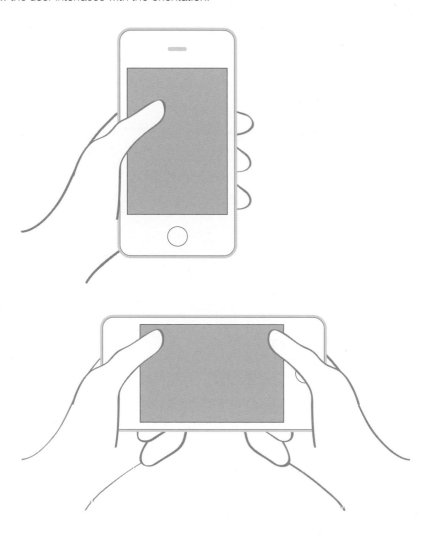

When a user changes from one orientation to another, how they physically hold the device will change. For instance, when the user is positioning the device vertically they will typically hold it in one hand, thus giving them the use of that hand (or just their thumbs) to touch navigation. In the horizontal orientation position the user will place both hands to engage the device, thus making it easier to explore, presenting them with more complex functionality and features. It also allows them to have one hand holding the device while the other is used to touch or swipe.

THE BASIC LAYOUT

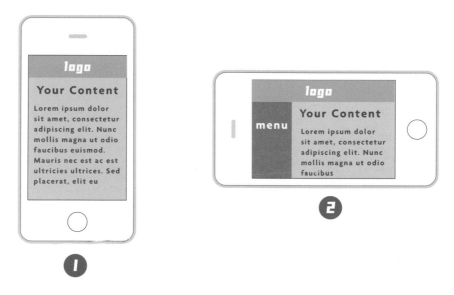

This pattern is made from two main components.

1. The Rotation (Gesture)

2. The Horizontal or Vertical Screen

Your mobile experience can be defaulted to one orientation, or it can change once the user opens the app or page.

The Rotation

This gesture acts as the catalyst to access your experience. Whether this gesture rotates to the left or to the right, the goal is to use this action to rotate the orientation of the screen. There is no recommendation as to whether the action should rotate to the left or right first; the outcome should always open the same screen, unless the design intent requires this to be different.

The Horizontal or Vertical Screen

Now that the screen is rotated, the screen can now change to fit the new formation. When working with horizontal screens you can provide the users with navigation to the left or the right of the screen that might have not been available when the screen was vertical. As the left hand would be holding the device, the user's right hand will be used to access the primary actions of the experience. This usage pattern is the norm for most right-handed users. (For left-handers, the pattern would switch to the opposite.)

When changing from a default horizontal view to a vertical one, the screen will be focused on a more compact view. I would suggest hiding navigation elements, allowing the screen to show a larger readable area.

THE ADVANCED ROTATE

The rotate pattern may seem simple, but it can open an endless range of possibilities to creating an engaging user experience. Imagine using the rotate function to create different views depending on the rotation angle of the device or using the rotation to open entirely new content. Let's take a look at both of these possibilities further.

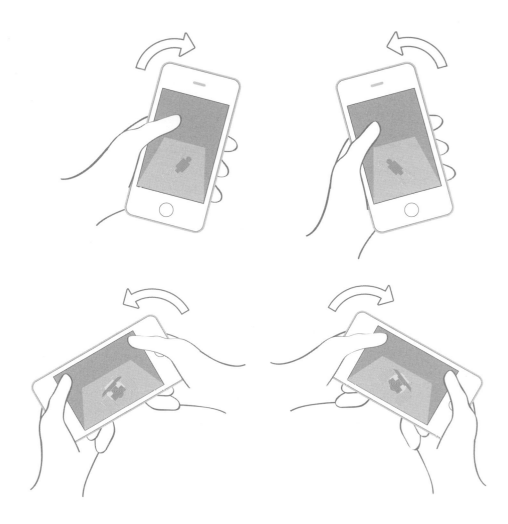

Designing for the Rotation and Tilt Gesture

The power of the accelerometer is that it gives us access to how the device is moving. Now let's harness that power to use it to help create an engaging mobile experience. The accelerometer is a perfect match to use as a gesture to control games. Using this gesture, our users can drive a virtual car by tilting left and right or in a side scrolling game use the tilt gesture to change a virtual characters direction from left to right. These are all great uses of these native controls to have your device act as a game controller for your app.

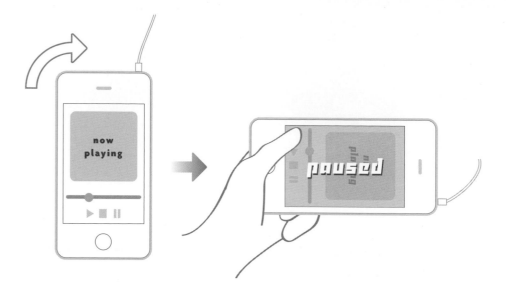

But let's take this a step further, imaging using the rotation and tilt to change navigation of our app or pages. Try using the tilt feature from left to right to access different navigation or functionality for the current screen. Why not make the user go back to on-screen navigation to change a view, open a new window, or access a different tool palette with each tilt or rotation.

A great use case would be a music player. As a complement to the on-screen controls you can have the user tilt to left to go back one song or tilt to the right to skip to the next song. The goal of this pattern is to use the hardware interaction to access and control the user interface. Designing your interactions should be tied to getting the best results for your users and providing an optimized user experience.

Changing the Content in Rotated Views

The next step would be use the full rotation of the device to change the on-screen content entirely. Instead of just changing the layout between horizontal or vertical, let's use this to change the screen from one view to another. Here are two possible use cases:

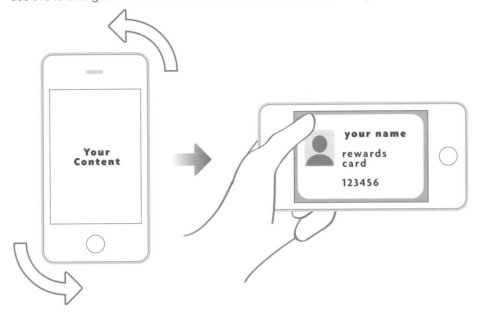

1. The Business Case

   The perfect case for this would be an app or mobile web page that requires you to have a user profile. Try using the rotation to access the user's profile. In one case of this, I used the rotation to show a user's membership card. This way, the users always had access to their mobile membership card at the flick of a wrist. Imagine using this to show your user's rewards card, frequent flyer card, or even a credit card. No more extra cards in their wallets or purses, and no more lost cards!

2. The Game Case

   Let's say you are in the middle of an intense game play session. You are about to step onto a subway car or hail a cab or meet a friend. Why not rotate your phone to pause the game? Instead of trying to stop and find the pause button you can easily pause the session with a quick flip. With the use of entire rotation of the device you know that your users wouldn't accidently do this during game play.

GOING EVEN FURTHER

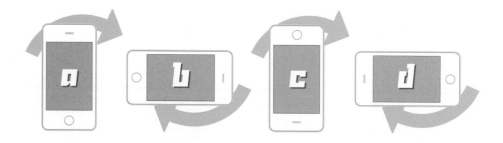

   To take this pattern even further, try having a different view appear as the users rotate their devices from top to left to right to bottom. Have fun with this pattern, but don't be annoying to your users. The goal is to experiment, but try matching the right pattern to the right user. For example, a user trying to write an email might not appreciate having their screen change when they accidently rotate or tilt their device; the gamer, on the other hand, might think this user interaction to be fun and engaging.

PATTERN 5: THE LOG IN

The login is your first introduction to how users interact with your mobile experience; deciding how and where your users log in is critical to keeping their experience successful. This pattern is a based on a flow of pages followed by a set of UI elements contained within each screen.

THE BASIC FLOW

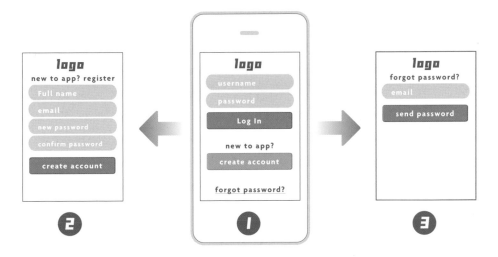

This pattern is created using three screens.

1. The Log In

2. The Sign-up

3. The Forgot Password

The Log In

The login screen itself is one of the most critical to any user experience. It has to allow a wide degree of functionality, yet be visually graceful; a tough challenge when dealing with a small screen size. The end goal of this screen is to make the process of logging in so easy to the user that they will not treat the login process like a road block to the experience.

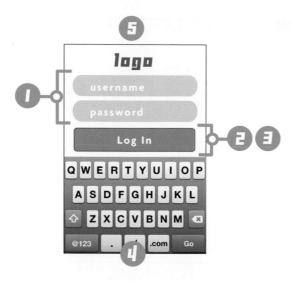

Let's break down a mobile login screen:

1. You will need to keep the input fields very simple. Remove any labels for these fields. (i.e., username, email, password, etc.); you can add these labels inside the field to save space and streamline the look.

2. Make the login button as large as possible. Its width can extend to the edges of the screen size. This way, you can make sure that it is large enough for a thumb to touch it comfortably.

3. Make the login button color very clear, and give it contrast to the background colors. The end goal of the screen is to make the user log in, so this action needs to be as unambiguous as possible. Make sure the color and shape of the button is distinctive and vibrant enough to catch the user's eye.

4. Design your screen to fit in the top half of the screen. As mobile devices display the keyboard on screen, you will need to account for the display area of the keyboard. As multiple devices and operating systems have different keyboards, you will need to account for enough space to handle these variations. At a minimum, the input fields should be displayed without having to make the user

scroll up or down. Imagine making the users scroll up and down when you have just cut the screen area in half; they will get lost in a moment!

iOS   android   windows 7

Login Screen with Different Keyboards.

5. Save passwords and user's credentials to the phone's internal memory. This can save time and frustration by allowing the user to return to our app or mobile website without having to go through the login screen. If we force them to go through the entire login process every time they return to our experience, they won't come back.

The Sign-Up

The ability to sign up a new user over a mobile device is critical to acquiring users. Launching your app or website without it is the same as having a party guest show up to your house with no ability to knock or open the door. Stranger things have happened;

see this screen shot of an iPhone app's log in from a well-known coffee company. This was the first screen when a user opened the app; what is missing? What do I, as a user, do if I am opening this for the first time?

Your sign-up screen needs to be concise and easy to fill out. Using much of the same guidance of the login screen, the sign-up screen should allow users to create an account with as few fields as necessary. I would recommend using an email as the username to even reduce the fields that the user would need to fill out.

Once a user registers, your experience should send your new user an email confirmation. This gives you a chance to interact with your user as well as assure them that the process of registering was successful.

The Forgotten Password

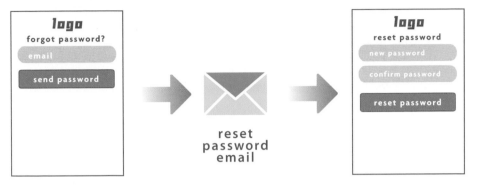

Not the most visible screen, but assisting with forgotten passwords is essential to completing the login pattern. The majority of all the work on this screen should be handled server side. The user should enter an email/username and have a password confirmation/reset link sent to them. This link will need to point to a page for the user to reset their password; this page needs to be mobile optimized, of course.

THE ADVANCED LOGIN: SOCIAL INTEGRATION

The next step in the login pattern will involve integrating social media credentials to log in to your mobile user experience. Why, you might ask? The premise is simple, if a user already has a Twitter or Facebook account, why not use their existing login credentials that they commonly use to log in your own app or mobile website. All of the common social media platforms include an Application Programming Interface (API) integration to allow this API call to integrate their login credentials into any website or app. As a designer, this gives you a few advantages:

1. Not reinventing the wheel—By using a social media "connect" call; you can streamline the process of signing up. This allows you to complement your existing sign-up and login screen from your mobile experience. The Connect calls open a screen to the social media platform, allowing the user to add their credentials. Once completed, the login process is complete; quick and painless. In some cases, you can use the social API to remove your login and sign-up altogether in favor of the social media login. Be aware that this will limit you to users who only have social media accounts. As a result of using the social credentials, you can gain greater engagement and monetization from those users who log in using their Facebook credentials.[1]

2. Completing your user's profile—One of the hardest aspects of creating mobile experiences is having users complete their profiles. Whether it's getting them to give you more information about themselves or having them trudge through filling out this information on a small screen; both can be frustrating to your users and you. By using a social media API, their existing social media user profiles can be passed to your app. Imagine collecting their full name, birthday, profile picture, phone numbers, and access to their friends/connections; all accessible through this login. This gives you a quick method of collecting more information about your user and still streamlining the login and sign-up process. For your users,

---

[1] Zoran Martinovic, "The Value of a Facebook-connected Mobile User," Facebook Developer Blog, January 24, 2013, https://developers.facebook.com/blog/post/2013/01/24/the-value-of-a-facebook-connected-mobile-user/.

this method gives them the ability to share and comment through your mobile experience to their existing social network and friends, thus providing them an incentive to register.

3. Making things secure—By using a social media API you will have access to a higher level of security that you might or might not have had access too. Better yet, you do not have to build or host this secure connection for your users' personal information or passwords. All of this can be handled by the API, a plus when you are trying to quickly build an app or mobile website.

AVOIDING MISTAKES

Example of a Poor Login Experience.

BEWARE OF SCREEN SIZE

You will lose screen real estate as you open the keyboard. As a result your login experience will become smaller than you realize. Focus on pushing the input fields to the top of the screen. Functionality such as allowing users to travel from input field to input field using next and previous buttons may help users, but it will be at the cost of decreasing the active screen area. Make sure your titles, inputs, and sign-in button are always above the keyboard screen.

BEWARE OF MAKING SCREEN ELEMENTS TOO SMALL

Yes, we want to make our screens as small as possible, but not at the cost of legibility. Titles and inputs fields should be clear and legible enough to read. These elements are just as important as your sign-up buttons as they will tell the user what they need to do on the page.

PATTERN 6: THE CAMERA

The camera has become one of the most important and popular components of a mobile device. Whether it's taking a picture on the run, texting a picture to a friend, or taking a quick video; having a camera in your user's phone or tablet gives an opportunity to build a mobile user experience around this feature. A perfect use case would be an app or mobile website based on video or photo social sharing, augmented reality, or mobile scrapbooking. This pattern is a new approach to using the camera as content creator. The navigation is based on using the camera as the main action in the experience.

THE BASIC FLOW

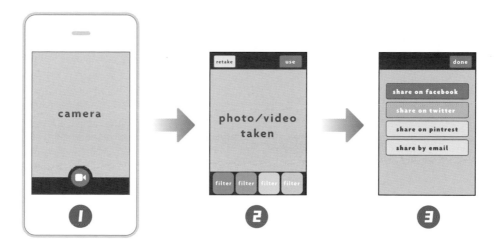

This pattern is created using three screens.

1. The Camera Screen

2. The Confirm/Add a Filter (Optional)

3. The Action/Share Screen

The Camera Screen

As the focus of this action is based on using the camera, the first item to be highlighted is the camera button. Unlike the default camera view, the ability of connecting to the camera using the phone's API allows us to create a custom visual interface to the screen. In our case, the action button is large and visible; to the user it is made clear that this is the next button that they need to click on. With access to the phone's camera, you can also set this action to use the regular camera or video by default.

Taking the design of this screen even further, you can add navigation elements at the bottom of the screen to access other functions. For example, you can add a button to access the camera roll or a user's profile. These functions should supplement the main task of using the camera function. Visually, these should be treated using secondary colors and be scaled smaller than the primary action button (i.e., the camera button).

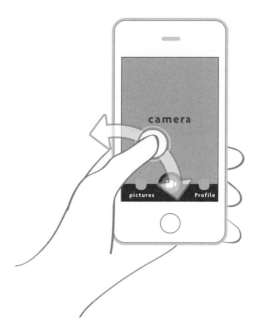

The navigation bar at the bottom also serves another function of expanding and contracting depending on different screen sizes. Plan on locking this element to the bottom of the screen. This ensures that the main navigation bar can change with the fluid screen size. *Why is it not locked to the top of the screen?* Locking it to the bottom of the screen has less to do with visual design, than range of the finger when the phone sits in a hand.

The Confirm/Add a Filter (Optional)

The second screen in the page flow is based on creating a method to confirm or retake the photo, or add a filter to affect the look of the photo to help clean, stylize, or color the image. The use of a filter can help personalize images for the user and it gives the experience a bit more depth than just taking a picture; the goal of personalizing the camera is our end goal for this pattern

The Action/Share Screen

The final screen in the flow is the action screen. This allows our user to focus on the desired action for the photo or video they have just taken. A typical case would be to share the image to social media platforms, email, send a text, or save the image to the user's custom camera roll. The intent here is to give functionality to what the user intends to do with the image. If the goal of the app is to socially share, then make the button for this action the most visible. If the goal is to scrapbook or make a travel log, then the actions provided should give an ability to save or add more details, like location, tags, or a description.

NEXT STEPS

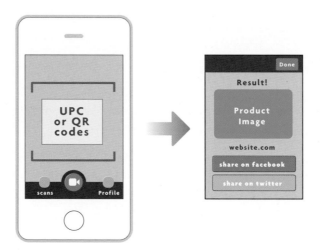

The camera pattern is a perfect example of using it as a starting point for building a more elaborate mobile user experience. For example, by adding a bar code scanning software development kit (SDK) to the experience a user can use this functionality to read bar codes or QR codes. The action screen can now be used to display, share, and save results from that scan.

WHAT'S A QR CODE?

A Quick Reference code or QR code for short refers to those pixelated black and white squares you may have seen recently. Think of them as an updated bar code for the twenty-first century. The power of the QR code is that they do not just store inventory information as a traditional bar code, but they have been expanded to store a variety of user-created data.

WHAT'S A QR CODE?—CONT'D

Here are some examples of QR codes you can create. Once read, they will display this information:

- URL (http://)

- Contact information (Vcard)

- Phone Number

- SMS Message

- Text Message

You can find several QR code generators online to create one for yourself. Use QR codes to provide links to download apps, share contact information, or provide a URL on printed literature. By downloading an app or adding a QR code reader API to your own mobile experience, you can read these QR codes. Your QR code will open the URL found in the code, display the text message on your phone, and download the contact to your address book. Used well, this can provide interaction with your users or potential users while they are on the go. Best of all, these QR codes work and can be read across all mobile platforms; sadly the reader does not work on the mobile web, only in apps.

PATTERN 7: THE MAP AND LOCATION

The common use case for a map application on a mobile device is to search for an address, location, or place. This pattern takes this interaction and creates more navigation elements to expand on the functionality of this user flow. This pattern can work for giving a user a method to search, preview, and access further details on these physical locations. Unlike using a maps application on your mobile device, using your own map interface can personalize the experience to your users and make each search result actionable. You can use this pattern to search and explore your own custom locations on Google/Bing maps or have the mobile experience load in your own custom map (i.e., a map of an attraction/event or floor plan of a building).

THE BASIC FLOW

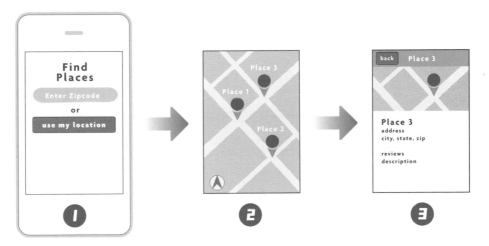

This pattern is created using three screens.

1. The Location Search (Optional)

2. The Map Page

3. The Location Detail Page

The Location Search

The basic location search screen acts as a front door to the user to present them with what they will be viewing and searching for. This gives you a method of communicating to the user this search action; for example, search for a hotel, a restaurant, a store, a museum, etc. This also gives them the clear option of using their location to define a search. Using their location will trigger most mobile devices' location services option. Below is a description on how location services are handled on the two common mobile platforms.

Android

Most Android devices use location services turned "on" by default. They use your location to periodically check the weather, time, and other services in the operating system. For Android apps the location is set by the developer when publishing the app and does not trigger any required action by the user. For the mobile browser the user sees a seamless integration to searching by their location.

iOS

On iOS, the request to check location will trigger the location services to provide the user with a notification. If they answer "yes" the location option is saved for your app whenever they access the app. If they answer "no," the location service is turned off for your app moving forward. Turning this option back "on" requires your user to manually turn on the location services from their settings screen. Once turned "off" you no longer have the ability to request this option; leaving your search by current location dead.

To best mitigate a user turning this off, we use a "use my location button" on the search screen to have them trigger this option. By having them trigger the option themselves they will be more likely to turn it on, especially if they are the ones doing the request.

The Map Page

The map page acts as a visual search results page for locations. Locations can be displayed with basic pins to mark their place on the map. Basic familiar gestures like the pinch to zoom helps users navigate around this screen. The functionality to choose "current location" should also be added to ensure that the user can use this tool to pinpoint places near their location.

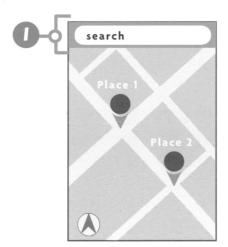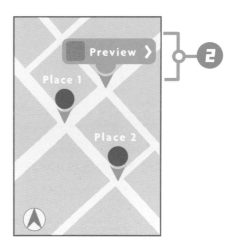

To enhance this experience optional UI elements can be added:

1. Search Bar

   A search bar can be added to collapse the search page onto one screen. This is helpful if the map page is part of a larger mobile experience. The search bar can also help to record previous searches by the user. The search bar will open the on-screen keyboard.

To make the act of previewing results even easier, a pin can be used to give a preview display to the user. I recommend adding more information like a location name, street address, and an image preview.

The Location Detail Page

The final page of this flow is the location detail page. The intent is to provide the user with more actionable details than a traditional map application. By adding UI elements to provide enough content and functionality, we can ensure that the user has enough substance to return to our experience over and over again. If you are going to use the map pattern as navigation for your experience, this screen will be essential to engaging your users.

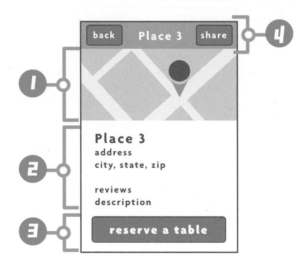

Here is a breakdown of UI elements on this page:

1. Map Preview

The preview will help to ground the user on the option that they have selected. This ensures that they have a visual cue back to the map page. I would keep this preview to a minimum of 20 to 30% of the page depending on the content you are showing.

2. Place/Location Description

   Make sure you have enough space to add context around the location, task, or place that you are asking the user to view. If you need more space you can integrate a list pattern to this page.

3. Primary Action Button

   In keeping with the mobile mantra of making mobile pages "actionable," you need to give the user enough direction to act on the results they have been given. If you asking them to search for restaurants close to them, then have them order a meal or reserve a table. If you are asking for a task, like checking in to their location, then provide a large "Check In" button here. Regardless, even a "Save to Favorites" button gives the user enough functionality to have them save their search results. Our goal is to not make them feel that they have gone through this process for nothing; include an action at the end!

4. Secondary Action Buttons

   Giving users more helpful tools or functions is never a bad idea. By adding these to the top navigation, it keeps the page clear for the primary action at the bottom of the screen. I would focus on giving the users access to tools or functionalities that enhance their experience.

For example, some functions can be:

- Sharing the location/place by email or social media

- Using the camera function to add pictures to this detail page

- Saving the page to the user's favorites

- Adding a rating to the page

WHAT'S NEXT?

    As you think about incorporating patterns into the design of your mobile experience, remember that your users are key. Ask yourself these questions. *What is the primary action you want your users to accomplish on a mobile device? What do you think your users can accomplish on mobile vs. desktop? What do you want their primary mobile experience to be on a smartphone or tablet?* By keeping these answers in mind you can focus your usage of the patterns to build the theme and direction for your mobile user experience. Use these patterns as a set of building blocks for you to build this experience. Remember that they can be tweaked, changed, and even combined. Take the map pattern and the list pattern and now you have an ability to expand the map details page. Take the launcher and the rotate and now you have the ability to show a user's profile without having to add an additional application to the launcher screen. Take the login pattern and the camera pattern and you have a camera application where users can log in using their social media credentials and share with their friends instantly.

    Some tips as you use the patterns to design your experience:

- Be direct! Don't have your users travel through multiple screens to get to the primary action.

- Design with intent. Make buttons look like buttons, make navigation look like navigation. Being subtle with your user experience design does not work for mobile users.

- Design with your users in mind. Focus on actions and interactions to fulfill the user's goals. Focus on the functionality they want and not something you think they need.

- Design with the device in mind. If you are designing around a specific device or operating system, make sure you keep all of those unique device characteristics in mind when you choose a pattern.

- Keep consistent navigation. Don't change navigation elements from one screen to another.

- Don't mix interaction models. If you use a gesture to open a menu, don't reuse that same gesture for something else on a different screen.

- Don't pack it in. Keep your mobile user experience lean and clean. Try to keep your mobile experience to a few direct actions.

- If it doesn't fit … If you can't get everything into your smartphone app, try giving the user more of the functionality you want in a tablet experience instead.

Don't be afraid to test some of these patterns in front of potential users as wireframes. Getting feedback on your design is critical to changing, editing, and optimizing the patterns. As we move forward, we will begin to discuss how best to display these mobile experiences onto different media to gather feedback. *That being said, on to mobile prototyping!*

CHAPTER 7

How to Prototype in Mobile

INTRODUCTION

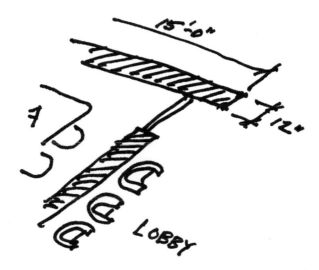

In a previous life (as an architect), I had an opportunity to lay out the interior of an office space. One of the most visible areas was a conference room next to the lobby; this was a perfect opportunity to design something that would connect both spaces. I sketched a floor plan and then an elevation drawing and had them drawn up. It looked great on paper and in drawings. I was excited to try out something unique that I had played with only on blueprints and never had a chance to build; here was my chance. Once construction was completed, I finally got to see my vision. In reality something was off and the design did not fit into the space well. This definitely wasn't something that could get fixed with paint. Ten years later that wall is still standing ….

Looking back, that project was a perfect opportunity for a prototype: to have built a scale model of the wall. Had I taken the extra time to create a prototype of my solution I could have saved a lot of cost in the long run and come up with a more appropriate solution for the space.

MOBILE MANTRA #3:
TEST, TEST, AND TEST AGAIN

WHAT IS MOBILE PROTOTYPING?

By definition prototyping is the process of creating a sample before a final version is made. In the case of the wall, a scale model of the space before construction would have sufficed. In the design of software, prototyping is used to create a version of the software that we can test drive and learn from. The end goal is to "work the kinks out" before a final version is made. Mobile prototyping is crucial to getting user experiences correct. Ideally, we want mobile prototyping to give us feedback on three aspects:

1. Page Flow—Does the flow of pages and screens make sense to the user? Is it easy to learn and navigate? Do we need to shorten or change the experience in any way? Does our user experience pattern fit into the users' expectations or conceptual model? We can use this feedback to optimize the overall user experience.

2. User Interface Interaction—Are our touch gestures clear? When we call a keyboard or picker from the device does the logic make sense? Are we using the UI element a user expects to see? With this feedback, we can focus on optimizing the mechanics of a specific screen, task, action, or function.

3. Device Interaction—What do our screens look like on the actual device? Is the use of the accelerometer clear? If you are rotating the screen, how does that look on the device? If you designed your experience for multiple screens, does it adapt well? We can use this feedback from actual devices to help us optimize the scale, proportion, and interaction of our experience with the device.

As we collect different types of feedback we can begin to revise the mobile user experience. My goal with this chapter is to introduce some methods of collecting that feedback. You will be introduced to different mediums that are available to you and how to use them for optimal feedback. As a result, plan on borrowing, editing, and using these to spark new ideas.

TOOL TIP

Aren't Using Wireframes Enough? Why Can't I Just Use Those?

Wireframes can take us only so far. In fact, be wary of showing these to clients or reviewers when you want feedback. Wireframes are *not* visual design; as a result, they might confuse or distract your reviewers from giving you the actual feedback you are looking for. For example, someone might ask why the mobile experience is in black and white, why there are no images on each screen, and why there are three screens on each page. If you do want to show your wireframe deck, make sure you preface the conversation with an explanation of what a wireframe is and what it is used for.

METHODS OF PROTOTYPING

Prototyping can come from several methods or mediums. Architects use 3D models, industrial designers use wax, and user interface designers use pixels. With mobile prototyping we will look at using a mix of paper and digital models to get results.

PAPER PROTOTYPING

Nothing says going "old school" like using paper to prototype a digital experience. As silly as it sounds, prototyping on paper can make it easier to gather quick feedback on the fly or to generate multiple ideas. It also sets the stage for team collaboration when sharing ideas. Remember, the goal of the exercise is to capture feedback.

The Mobile Sheet

The idea behind the mobile sheet to is place one screen on a single sheet of paper. Sounds simple enough right? Imagine what that single sheet can give you. That sheet gives you real estate to write down feedback as someone gives it to you, enough space to sketch or draw an idea you get while you review the sheet, the ability to arrange and shuffle the sheets as need be. With that single sheet you have the power to throw away or hide a screen at a moment's notice. Now imagine you are showing your experience to a group of people in one day, you have a collection of numerous sheets per person and per screen. Staple them together and now you have a wealth of knowledge and feedback for your user experience. Not too bad for a single sheet of paper.

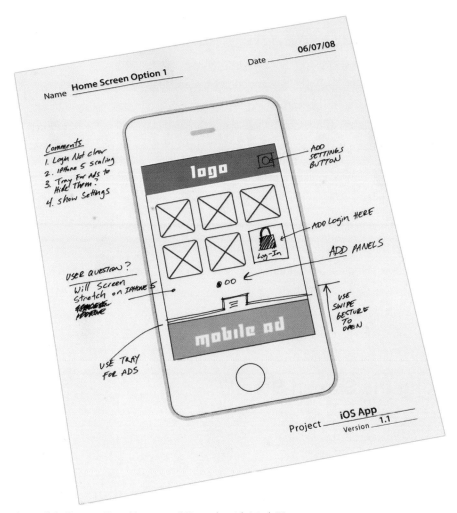

The Mobile Sheet—Clean Version and Example with Mark Ups.

Let's start creating your own. First, choose a page size. I recommend using a standard letter size (8½"×11" or A4); this makes it easy to photocopy or print if you need quick copies. Second, choose a device skin for your mobile sheet, use one that matches the right screen size and project type. Third, place the device skin and screen onto the page. Give yourself enough space if you are planning to draw or take notes in the border. Now your mobile sheet is ready to add your layouts!

As you add your layouts for your mobile user experience, here are more recommendations. Use black and white for your layouts; this makes it easy to reproduce. When you do print, make sure to print out some empty layouts to add or rework screens later. You will likely encounter the scenario of someone saying … "I wish the screen could do something like this. …" You can quickly sketch out the screen and respond with "does it look something like this? …" (This is a crowd pleaser.)

Now that you are done printing, make sure to use a paper or binder clip to hold them together, you want the ability to shuffle the sheets around, if need be.

But what if you want to show more interaction? This is a mobile paper prototype, isn't it? One of the great thing about paper prototyping is its flexibility … no seriously, that's not a joke. The great thing about paper is you can cut, tear, rip, and tape together if you need to without feeling like you have to formally move screen elements around or rebuild your UX. Changing and editing is instant. Here are some exercises you can use to help.

Making the Screen Scroll

WHAT YOU NEED BEFORE YOU BEGIN

- Exact-o blade or utility knife

- Legal or long sheet of paper

Concerned that you are not showing the screen scrolling on the mobile sheet and want to demonstrate this? Try this little trick.

1. Cut the top and bottom of the screen area; use an Exact-o blade or utility knife to get a nice clean cut.

2. Print out your entire long screen on a legal-sized paper (8½"×14").

3. Fold and trim to make sure it matches the width of the cut you made.

4. Insert the folded paper into the two cuts with the screen side facing up.

Cutting Top and Bottom of Mobile Sheet.

Slide the screen up and down. In a few steps you can quickly show the scroll of the screen in a mobile sheet. It's simple and quick, but very powerful.

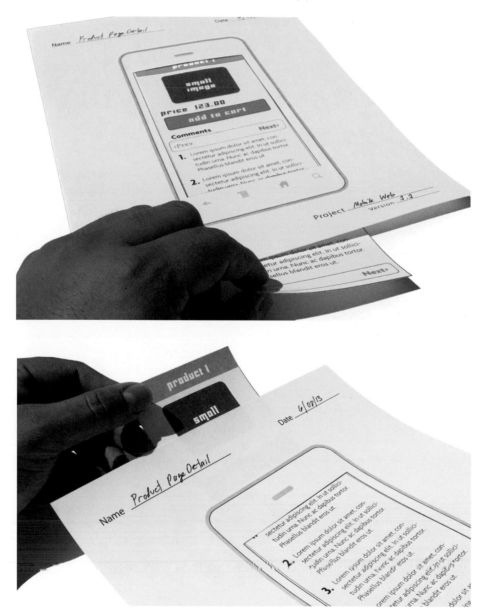

Running Screen through the Cut Mobile Sheet.

Showing Inputs, Keyboards, and UI

Do you want to show a keyboard or menu overlay on the screen? Try this next exercise. You will need to do some prep work before you begin.

1. Scale and print several of your UI elements before you begin. Make sure they are the same width and scale as your mobile sheet.

2. Once printed, cut them out with scissors. Make sure you cut out some extras.

3. Add a piece of tape to each paper UI element.

4. Attach each UI element in its appropriate location and fold over face down.

Cutting Paper Inputs.

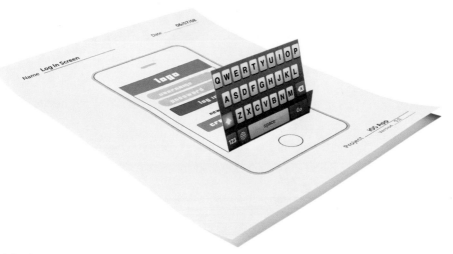

Mobile Sheet with Paper Inputs.

Once you have this ready, you can begin to experiment folding over each paper UI element. If you want to show how and where the keyboard folds over on the screen, simply fold over your paper keyboard. If you want to show how your menu comes in from a tray to the left of the screen; fold over the tray from the left. This is a fast and quick way to show how a screen changes without shuffling through numerous pages. Even better, if you find yourself with a user who thinks you need a different keyboard or picker; grab from your paper cutouts of UI elements and tape over it. If you find a user who thinks the tray should come in from the right, then remove it from the left and reattach on the right! This is the quick way to change your navigation or UI in an instant.

TOOL TIP

These Paper Prototyping Methods Are Great! Can I Use These Mobile Sheets for Creating Wireframing as Well?

The mobile sheet gives us the ability to draw and sketch out ideas on them as well. Print out a clean copy to use as a base for sketching wireframes of screens and functionality. For sketching out wireframes, I recommend using a wireframe sheet with the three-screen flow. You can also use the cut out inputs to add to your wireframe sketches. This method is perfect for collaborating with other designers or developers on how a screen may or may not work. Make sure you date each sheet when you are finished.

For sample mobile sheets see Appendix C.

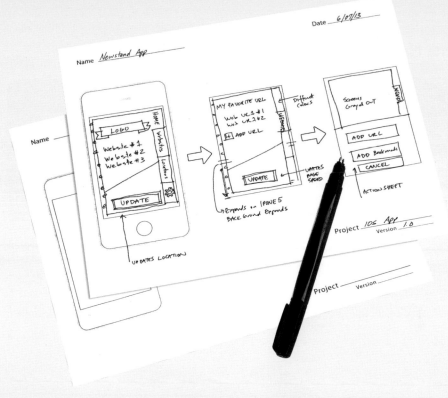

Mobile Sheet Example #1 for Wireframing

Mobile Sheet Example #2 for Wireframing.

The Notecard

One of my favorite paper prototyping exercises is the use of the notecard. One of the reasons for this is that it allows for the most collaboration when laying out or editing a mobile user experience. This also takes some prep work, but the feedback you will get from it pays off tenfold. Plan on setting up several group exercises to share, collaborate,

Mobile Notecards.

and get feedback on your user experience. I will walk you through some of these exercises so you can set them up on your own.

Before you start any group exercises you will need to do some preparation. Start by transferring all of your screens onto the notecards. Either draw them on each notecard or tape a copy of your printed out screens. You could also just print out your screens on regular paper, but I recommend using a heavier paperweight to be able to shuffle and handle these screens. Your mobile notecards might get bent or folded if you are conducting an exercise with several participants. I recommend making two or three copies of your mobile user experience notecards.

EXERCISE 1—GROUP

Find a room with a nice wall or board you don't mind punching pinholes into. You can use tape, but from my experience notecards don't always stick well. On your wall arrange your screens from left to right as shown in your wireframes. Make sure every row represents each unique page or function in your mobile experience. If this activity looks and sounds familiar it's because it closely resembles storyboarding sessions for animation.

Example of a Storyboard Layout of Notecards.

Once you have them up, review andwalk through each function or navigation path with the group. If something doesn't make sense, feel free to remove or rearrange the notecards. The goal of this exercise is to see the entirety of the mobile experience. How does a user first enter? Are the interactions consistent? Is the navigation path consistent? Is the mobile experience too much? Ask questions like these to spark the group conversation. If you want to allow for more collaboration and activity, have the group members walk through the experience themselves. Use sticky notes to add annotations or mark areas of interest or problems. Once you're done BE SURE TO PHOTOGRAPH THE RESULTS; yes, this sentence is in all caps. Document the results of the session, once the notecards are down, you will lose all of your feedback and any changes to screen position or edits. If you are lucky enough to have a wall that you can keep your cards up on, then this is helpful. Even though these might get changed, always document your progress.

EXERCISE 2—INDIVIDUALS

Mobile Notecards Scaled for an On-the-Go Testing.

The next exercise is focused on using your stack of notecards to show your mobile user experience to an individual. In this case you will be using your notecards like a deck of playing cards. Before the reviewer comes in, organize the deck of UX screens in the order you want to present them. You can then present your mobile experience one screen at a time. You can also lay out the screens in order on a table. The great thing about this notecard prototype deck is you can present it anywhere (on a train, in a park, or on the go)! In fact, the size of the notecards and their portability is very reminiscent of a certain mobile device you are trying to emulate; see my point? If want to take it one step further you can even scale and trim the notecards to match the size of a mobile screen.

DESKTOP PROTOTYPING

The next step to paper prototyping is doing so on your desktop. There are no hard and fast rules here. I will show you two methods that may help with presenting your mobile experience wireframes. These will include a simple and an advanced method.

The Simple Method

Use whatever presentation software you feel most comfortable with to lay out each screen on a single page. Once you have done so you can use the presentation mode to act as a click through. Simple enough? All you really need is a simple method to create a walk through for your wireframes. If you want to add a device skin to give the screens some content, go ahead. Whatever you choose, try to adhere to these rules.

1. Is it simple?—Choose software or a file type that is simple and easy enough to move or edit a screen.

2. Is it portable?—Choose software or a file type that is portable. Make sure you can quickly email or send it to someone and have him or her open it.

Remember that you are not trying to replace your wireframing tool. Most wireframing tools have a presentation mode as well. I recommend keeping a separate format or method just for presenting these screens. You don't want to have someone tinker with your designs as you try to present them. Use Microsoft PowerPoint or Apple Keynote as your simple presentation method or to create PDFs to email.

The Advanced Method

If you want to take this one step further, there are an infinite number of software packages that allow you to wireframe and then build a clickable prototype. Most wireframing packages like Omnigraffle or Axure give you the ability to link pages from within each screen. Just do a search on the web for "desktop mobile prototyping" and you will find a slew of web-based wireframe/prototyping packages. Choosing a package is like choosing a flavor of ice cream, everyone has a favorite and everyone will have an opinion. The choice will be up to you.

As you explore choices, keep in mind the two rules of thumb from the simple method: is it simpleand is it portable? I love the idea of web-based packages, but then you need web access when you present. I love the idea of using a software package I already own and do not need to learn, but then I cannot send a presentation to anyone. I love the idea of sending someone a pdf, but then it's not really interactive. Here is an anecdote from my own experience.

Screen Capture of My Actual Flash Prototype.

A long time ago (a long time in mobile years was 2010), I worked on a project that required us to do a prototype of our mobile user experience. The client required that we be able to send this prototype to be reviewed by different business teams and get final approval from a legal team in another city. Would we have to send them an iPhone or mockup an app? This was a dilemma. On a whim I built a quick prototype in Adobe Flash. This was a great ongoing joke with the team as I had just built a virtual iPhone on a platform that it didn't support. I even added the functionality of the home button to turn the screen on and off and a space to add a clickable app icon as well. Once we had a good laugh, we realized the power of this prototype. It was simple enough to drag and drop in a screen, edit a frame to remove a screen, and even export it as an application and send it over email; no special software required! This was a hit. A few years later, I talked to someone who told me that the flash prototype was still floating around at that company.

Looking back with what I know now, here is how I would have changed my approach and what I do to mock up my own user experiences today. For starters I would go the route of using a simple method to present the screens. PowerPoint can be your friend as everyone has a copy (or access to one). Second, as most people now have a smartphone, I skip the desktop prototyping all together. The more I can show on the device the better.

TOOL TIP

Why don't you use the SDK iOS simulator or Android emulator on the desktop for prototyping?

All mobile SDKs provide an emulator for testing and debugging of apps and code. In order to use them you will need to download, install, and setup a development environment for each mobile platform. These tools give you a very limited view of the operating system and include only a basic set of functions. They only allow you to install your own apps directly from the SDK and do not allow any outside apps to be installed. They include access to the browser and some methods of activating on-screen gestures (i.e.,, in the iOS simulator, the shake can be triggered from the desktop menu).

Using these tools is great in theory, but you are limited when it comes to using them to show a prototype. You can easily run your compiled app from the SDK, but cannot load in your own screen shots or images to look at from inside the emulator. In order to do so, you will need to create a web page to view images or pages and open them through the emulator's browser. These tools do not include an offline mode so you must be connected to the web from your desktop to open these pages. At that point it is easier to open them up directly on your own device. The emulator will also not have the same performance compared to an actual device. For example, the Android emulator is notoriously slow even after several upgrades over the years.

When I plan to test a prototype, app, or mobile website I skip the emulators and use a physical device first.

USING THE DEVICE TO PROTOTYPE

As you remember, Mobile Mantra #2 is Be on the device! Once I am on the device I can get feedback on three things that paper or desktop prototyping will never give me:

1. The correct size and proportion once it is on the device.

2. The correct colors and feel of the screen.

3. The feel of the device as it sits in my hand and as I look at the screen.

Here are the two prototyping methods that you can use on your own devices.

THE IMAGE PROTOTYPE

This method is the simplest, most time effective, and packs the biggest punch. Start by taking your screens and scaling them to the right screen size; for example, if you intend to show them on your iPhone 4 scale them to 640×960 pixels. Save them as PNG images so they do not have compression. Next, send them to your smartphone over email or by any other method you use to get images to your phone (cloud storage, text, etc.). Finally, save them to your camera roll and arrange them as you want. Now, walk around like you own the place!

Image of Wireframe Image on iPhone.

Image of the Same Wireframe Image on Android.

Congratulations! You have just put your wireframes on your mobile device. Sounds too good to be true? Yes it is. As simple as this is, I cannot tell you the times that I have used this simple method to show off screens. You can quickly swipe from screen to screen and get quick feedback on what a screen looks like. Want to edit or change a screen? Just email yourself a new screen and replace it. Want to test it on Android? Just send it to an Android device! You can even use your mobile device to record audio of your users' feedback. Again, this method is perfect for gathering quick feedback on a design, for anything more complicated and clickable look at creating a clickable prototype.

CREATING A CLICKABLE PROTOTYPE

Want to create something more interactive? I suggest creating an HTML5 version of a click through prototype. It's easy to maintain, editable, and best of all, it's free! Here is how you start. Using an html editor to create an html page and make sure to set the width to the screens you will be testing on (tip: HTML5 does not require you to set the doc type to mobile). Don't over think or over do this. Use simple linking to connect pages together. Your goal is to create a simple click through and give the user a basic navigation path. Want to add more depth? Add in clickable areas to allow the user to explore pages. Or add some simple fields to allow you to launch the keyboard or dialer. Once you are done, uploaded them onto a web server. Now you can load them onto your device's browser; bookmark these pages to make them easier to open multiple times. One of the advantages of this method is that you can use these prototypes to gauge the performance of these pages while on the device (see Chapter 9 to learn more).

TOOL TIP

What if I want to create something to show gestures and touch events?
Do a quick search on the iOS or Android markets and you will find several apps designed for the sole purpose of creating a clickable prototype. As a result, a whole new series of apps are now available to download; some free, others you will need to purchase. If you are building apps or mobile websites for mobile, why not create your prototype on the actual device? You can also use these same apps to create wireframes as well. I have collected just a few here: www.mobileuxpatterns.com/prototyping.

As you work on your clickable prototype of choice, keep this in the back of your mind; beware *the prototype trap*.

The Prototype Trap

As you build your clickable prototype you will begin to think to yourself, "What if I add just one more function, just one more click, just one more page." In fact you are starting down the steep road that most prototypes travel down ... the downward spiral of the prototype trap. Many have fallen prey to this scenario. Here are some hints that show you are travelling down that road. Is your prototype so large that it requires hours to edit? Does it require several people to maintain? Do you have as much functionality in it as the final product does? Do you have to add release numbers to your prototype? Did your mobile developer say, "I can build a prototype in the Android or iPhone SDK"? Is it over 100 MBs? If you answered yes to any of these, then you are in the danger zone, time to take a step back. A good prototype should be made fast and repeatable. Translation: it should be disposable. The prototype is about the feedback and not the technology or the method that made it. For mobile user experience especially, you should be able to produce one quickly, gather feedback from a user, and be able to change it. If it is on the device, even better. This needs to be a fluid cycle, like the mobile, it will always be changing.

CHAPTER 8

Mobile App or Mobile Web: The Big Debate

INTRODUCTION

Choosing to build either an app or a mobile website is one of the biggest debates. Like choosing a political candidate, each opposing side can get verbal, adamant, and emotional about their choice. Everyone you talk to will have an opinion. It might come as a friendly anecdote or as a stern rant; one way or another, you will hear every side from every different angle—if you haven't already. I was once quoted as saying: "We are in the middle of a love-fest with mobile applications. Every brand feels the need to develop one, believing this is a mobile panacea—the easy route to be relevant, accessible, and cool."[1]

I will present you with advantages and disadvantages of both apps and mobile websites. An overview of each will help you make an informed decision; choosing the one you want to build will be up to you.

THE MOBILE APP

One of the *great* things about the mobile app is that it created a buzz around the mobile, yet one of the *bad* things about mobile apps is that it created a buzz around the mobile. On three separate occasions I have had random people say to me, "Have you heard about these things call apps? I hear I can make money selling them...." (To be more precise about from whom I heard this: my mailman, a cab driver, my mortgage

---

[1] Adrian Mendoza, "Why a mobile app does not make sense," *Mobile Marketer*, December 27, 2010. http://www.mobilemarketer.com/cms/opinion/columns/8605.html.

broker, and a salesperson at Best Buy.) It felt like each of them was trying to give me a stock tip. You can't watch your favorite television show, read a magazine, or surf a website without being bombarded with an advertisement about an app: why you should be downloading it, why you need to own it, or why it's better than another app. Putting the hype aside, let's look at what building a mobile user experience for an app will mean to you.

ADVANTAGES
Built Using Native Code

One of the main advantages is that apps are built using "native code"; this means an app is built in a programming language that smartphones use to run other existing apps and its operating system. For iOS it's Objective C and for Android it's Java. As a result, your app gets access to using the processor and memory allocation. What does this mean to you? This means you can use the power of the phone's processor and memory to run heavy calculations or functionality; for example, use an app to run a 3d game, rendering dynamic pages and running tools that require processing. As apps are installed on a device they do not require a network connection to open. You can read content, play a game, and interact with an app without being connected to the Internet; a plus when flying on a plane or travelling in areas with no signal or low signal strength.

Using Device Specific Features

Aside from using the processor and memory, you can also have easy access to some of the device specific features on smartphones. The app gives you access to the camera, accelerometer, GPS location, camera roll, contacts, and the different keypads/keyboards; all of which you can use when designing your mobile experience. Aside from the UI components, one of the other features you can design for is the use of app notifications. You can use notifications to remind, alarm, or even update your users about news, events, and updates; a great feature when you want to engage more with your user or promote more interaction with the app.

Design for Known Devices (iOS)

One of the features of working with an app Software Development Kit (SDK) is the power to build an app that will work seamlessly with the devices you are familiar with. This comes in the form of native code, but also in the framework for adding in UI features. The iOS SDK gives you placeholders for adding in all the icons, backgrounds, splash screens, and creating a presentation layer for building your app for an iPad app as well. It also gives you access to using their library of UI elements for the iPhone and iPad. This alone is a timesaver, by not requiring you to design a whole new set of UI elements and interactions.

DISADVANTAGES
Design for Unknown Devices (Android)

One of the problems with Android is the wide variety of screen sizes and proportions. This problem grows exponentially when you have to build a native app that has to respond to all possible cases. This process becomes a logistical nightmare as you have to cut graphics and icons for every possible case you encounter. Not only do you have to refactor your UI for these cases, but you also have to test on every different OS version, manufacturer skin/flavor, and device. This is a problem that will consume your time when designing and building your experience. You need to account for this as part of your apps testing strategy; make sure you have also accounted for the time that it will take.

Pay to Play

On iOS you will need to pay a yearly fee to have access to the iOS SDK. Upon your first payment you will be able to download the iOS SDK to start working with software program Xcode. More importantly, you have the ability to build on your actual device and publish to the Apple app store.

On Android, all the necessary software is free and can be downloaded online. It uses the software program, Eclipse for Android development; existing Eclipse users only need to install the Android SDK and they are ready to go. This is a plus as most software developers already have Eclipse installed. Having access to the Google Play store will cost you a small fee for your Google account to have rights to publish on their app store. With this flexibility you can publish different apps from multiple Gmail accounts; not a bad idea if you want to publish app as different brands. Each Gmail account will be charged that small fee. So yes, you will have to pay to Play … no joke.

The App Store

Once you have built your app and tested it, now you have to deal with the app stores themselves. This is a much larger task than it seems.

- The app approval process on iOS—the dreaded approval process that terrifies children and is the topic of scary campfire stories. … Truthfully, if you design to the human factor guidelines, you will be fine. The scary part is not the approval process, it's actually the inconsistency of it. Give yourself a month to get your app approved. In most cases, if you push an update after your first release it may get approved faster than your first release. With no guarantee on how long it will take for your app to get approved and no documentation on the process for approval onto the app store you will need to give yourself ample time for this review and approval process (schedule about a month, if not more).

- The app approval process on Android—well there is NONE. Anyone can post an app; it takes 15 minutes to add a description and some images. Why is this a disadvantage? If anyone can post up an app, then anyone can impersonate you or your company. With no approval or review process it's the wild, wild west of app distribution. By having this process too open, many malware and viruses have made it into android apps on Google Play posing as legitimate apps.

- Discoverability—Apple and Google advertise that they have millions of apps on their app store. If you have a game or great app that you want to get noticed, then you are also competing with a few million other people who want to do the same. Trying to raise your discoverability in the app store is considered a black art; no documentation or guidelines exist.

- Monetization—Want to sell your app for $1? You need know that Apple or Google will take a percentage of your sale, 30% for every dollar you make selling your app. Planning to sell subscriptions, newsstand magazines, or to support an app purchased? The same economics apply.

MOBILE MISCONCEPTIONS FOR MOBILE APPS

1. You will make a pile of money selling apps—In reality, no one will become a millionaire selling apps. Most developers will never make a dime or break even. The mobile app has become the new gold rush of the 21st century. And, like the gold rush, there are booms and busts. Focus on designing an app that fulfills a purpose and need for users, not on the idea of selling it to make a dollar.

2. Making an app is easy—Making an app is building a piece of software. Think of it as that analogy, because it is true. If you are a nontechnical designer, most SDKs will not make sense … remember it's a complicated piece of interactive software you are building here.

THE MOBILE WEB

The mobile web has long been considered a backup solution for mobile designers. It has long been overshadowed by the mobile app. One of the big reasons was the fact that the first mobile websites looked more like websites from 1995 than a modern-day mobile experience fit for a smartphone. The mobile web has long suffered from competing with sexy slick apps and the technology limitations of html. But, the story of the mobile web is not over yet; its popularity is now increasing with the rise of HTML5 and responsive web design. The next chapter of the mobile web is just starting.

ADVANTAGES
Everyone Has a Browser

The key advantage of the mobile web is that every smartphone has a browser installed. With the browser there is no need to download, install, or register to an app store. There is no need to create different applications for operating systems; your mobile website will open across all devices, browsers, countries, and operating system versions. By using the browser you can have you mobile website open from other websites, open through a link in an email, and open through a mobile ad.

No Special Code to Build

Building your experience for the mobile web requires no special programs to install or SDK to buy, it is based on the good old html code. It is easy to change and much easier to find someone who can build your experience for you. HTML5 in combination with CSS 3.0 gives us the use of elastic layouts and some new and improved aesthetic features for us to use (transparencies, gradients, drop shadows, and more).

The use of HTML5 also gives us the access to APIs (Application Programming Interface) that allows us to enable a smartphone's specific feature. We can now call a specific keyboard (i.e., web keyboard, number keyboard, available to iOS, but limited in Android) and the dialer. It also gives us access to load page components or elements into the phone's local memory. We can use this feature to download heavy page components into the phone's cache. This gives your user the ability to read content offline and to not have to download your mobile site every time they visit; improving the performance and paving the way for the creation of more elaborate experiences. This new fusion of software and website is referred to as a "web app." Want to provide geo-location data for your user to use in their web experience? Now you can! You can design your experience around accessing the phone geo-location data and use functions to request the smartphone's current location as well.

No Need to Push Updates

In the world of apps that require an app review process and an app update, the mobile web is updated when and how you want. Your mobile site is always updated and each user always gets the most recent version. As a result, you can update the mobile web with small changes on a regular basis; compared to an app. This is a huge cost and time saver compared to the time needed to push app updates through the review process.

Monetization Is All You

The best thing about the mobile web is that access is free. This allows you to sell subscriptions and access to your paid mobile content without giving a percentage of profits to anyone. This alone has caused some online magazines, like *The Financial Times*, to switch from apps to mobile websites. By using the feature that saves data to the device's local cache, this allows their mobile site to download content to the user's phone from offline reading—the power of a "web app!"

DISADVANTAGES
Redirects Are Still a Problem

One of the main problems with the mobile web is defining the redirects from your desktop website to the mobile version. A redirect is a piece of code on your website that allows you to change the flow of traffic and point to a mobile version based on the incoming device's operating system. Some redirects allow you to use a user-agent string[2] from the device to make this choice while others use JavaScript to resize the page; both require you to keep constantly updating this method as new devices and screen sizes get introduced. If you want to redirect only iPhone 5 users to a different experience, with the current methods available this will be a difficult task. As you can tell there is a lack of consistent format and method to redirect your traffic to different mobile experiences.

---

[2] User-agent string—value from the device that describes the type of device and browser it uses.

Multiple Mobile Sites

Regardless of the redirect, you will still need to define how many different mobile websites you will need to build and design for. Will you create a *m.website* for your mobile experience (no standard dictates if it's a *m.website.com* or *mobile.website.com*)? Will you create a site for iPhone and Android separately (*iphone.website.com* versus *android. website.com*)? And now add the next level if you want to add a tablet website. Does the tablet redirect to your desktop experience or do you create a *t.website*? Imagine having to do one change to the UI and then having to replicate that change over and over again on all of these different versions. As you can tell, juggling all of these different versions, user experiences, and code bases creates a logistics nightmare for you.

Not All Mobile Only Features Are Available as APIs

HTML5 has opened up a whole new world by allowing us to use APIs to connect to the mobile device, but at the same time, new API features have just started to trickle in. While apps are already using device-specific features like the accelerometer, rotating the device, use of the camera, and access to the camera roll for some time, mobile web experiences are still catching up. Even though some of these APIs have been available to the desktop web for some time (geo-location, camera), they do not consistently work across all mobile platforms and browsers. There are amazing features to use in your mobile website, but they require you to do more work in researching their compatibility before designing them into your mobile experience.

RESPONSIVE WEB DESIGN

Responsive web design has given the mobile web its second wind. It has created a new excitement and helped spark new conversations about designing for the mobile web again. The premise is simple; use HTML5 and CSS to create elastic and flexible layouts for your web design. Using CSS, you can set the sizes and width to scale and shrink web pages. The same website that is your desktop site can become your mobile site and then your tablet site. One code base, an optimized experience for each device size.

ADVANTAGES
Say Goodbye to Redirects and Say Hello to Break Points

Responsive web design (RWD) uses what is called "break points"[3] to set widths for each screen size. When a browser or device senses the screen width, the RWD page will scale itself to the set width. This allows you to create a smartphone, tablet, desktop, and other experiences using the same web page. The goal is use break points to create a fixed set of widths, start by creating a smartphone and desktop width first; any widths in between are scaled using elastic and flexible page elements. This allows you to scale your web page to create multiple different screen size experiences, letting the browser do all of the heavy lifting for you.

No Need to Learn Another Piece of Software

The power of RWD is that it uses existing code in HTML5 and CSS 3.0. There is no need to learn a new language or download software. Like using traditional HTML for the mobile web, all you need is to learn the new break point patterns, and away you go. Building a simple page that can scale around multiple break points can be done very quickly.

DISADVANTAGES
RWD = More Work at Your End

The power of RWD is also its weakness. Adding multiple break points is great in theory, but will now require you to change how you design for those widths. Remember that not every experience will fit perfectly, so you will have to tailor and optimize your experience as you build your website. Taking this into account, here are three things you will need to change.

---

[3] Break Points are also referred to as media queries, look at http://mediaqueri.es/ for examples.

1. Design for components and not screen widths—When developing RWD pages, all of the page elements and components you design will have to be flexible as well. Forget fixed widths, fixed layouts, and unique UI elements that are constrained; every element on the screen will have to be designed in percentages and flexible widths. Be prepared to wireframe and describe what each individual component will look like as its scales.

2. Change your team's thinking and process—The process and workflow that your team uses will have to change. Gone are the days of a waterfall development schedule, where individuals finish their work right after each other. To get your RWD project working correctly you will need to tie your development, design, and user experience team closely together. This will force a tight collaboration that is necessary to catch all of the changes that arise from trying to match and optimize screen widths. Forget about having a separate mobile and desktop team; both will need to be working and thinking on the same page. The hardest part of this will not be developing the RWD site, but it will be convincing your team to change their thinking and process of how they approach a RWD project.

3. User experience design needs to evolve—Gone are the days of designing a set of wireframes and walking away. When working on a RWD project, expect to be heavily involved throughout the entire project. As break points and screen widths are being built you will need to come back and redesign wireframes and workflows. As pages get scaled, you will need to wireframe page components in different widths. Your user experience design process will need to become more interactive and require you to design on the fly as changes happen. As you design more complex RWD experiences, it will require you to do more complex planning at your end; so give yourself more time and budget.

Rethink Your Fancy Web Design

RWD creates a scenario that limits some of your design choices. Because you will need to create a design that works across all different widths, you have to design to a specification that will work regardless of the width. I have had two such experiences that I would like to share.

1. I worked on a RWD website that was designed to use the Parallax effect; an effect in web design that layers images and elements over each other as the screen scrolls. This fancy effect was perfect for the design and user experience of this particular website. When planning the development and user experience it appeared like it would be very difficult to add. This effect does not work properly on most smartphone browsers or even tablets. As a result it came down to a simple design decision, choose the parallax effect or RWD, but not both.

2. The next example of a RWD site included a nice java script carousel. It became apparent that the carousel we liked did not scale well and neither did the AJAX script that ran it. RWD will require you to rebuild and upgrade all of your existing AJAX tools and code libraries. In some cases, code and dynamic elements you once used on the desktop will not translate well to the new mobile web experience.

Performance Problems—The Double-Edged Sword

The last disadvantage and most important one, is what I refer to as the "double-edged sword" of responsive design. If you are using the same website for the desktop, the tablet, and mobile, wouldn't the actual size of the website be exactly the same? If you guessed *YES* you are right. RWD opens exactly the same website in all three scenarios. For example, if the size of your website (including JavaScript, images, CSS, and html) is 5mb (extreme case) it will open the same size website both on a mobile device and a tablet. Not a big deal if you are on your desktop connected to Wi-Fi, but this is a huge impediment to your mobile users who are on a 3G connection. Why is this critical? If your website opens in 5 seconds on your desktop, expect it to open over 15 seconds on your mobile device. At that point you will have some very unhappy or nonexistent users.

Fixes exist to optimize for mobile, but these require more work at your end.[4] Tricks like compressing your CSS, adding java script to change your image sizes,[5] and stripping out any images in favor of CSS are different routes you can take. Regardless of any tricks, hacks, or new code you add; this requires you to add another layer of complexity onto your mobile experience. The great solution that would have worked across all platforms and sizes now adds more work on planning, designing, and even coding.

MOBILE MISCONCEPTIONS FOR THE MOBILE WEB

1. It's small, so it's easier—Think again. The mobile browser is less forgiving. Mobile websites may be smaller, but they require more planning to get them right.

2. Responsive design will fix all my problems—RWD works when you can hit the reset button and create a brand new experience from scratch. Trying to force it into your current desktop or mobile experience is difficult. At times it will be easier to start from scratch, but not everyone has this luxury.

THE TRIPLE PLAY

I first heard the phrase "the triple play" to describe Bank of America's mobile strategy. The triple-play strategy provides their customers access to online banking through their mobile app, mobile websites, and SMS text messaging. One way or another, their services will be available to any of their customers' different mobile choices.

---

[4] Interview with Jason Grigsby, Cloud Four April 30th 2013. — Quote: "It's your job, Suck it up!"
[5] Jason Grigsby, "8 Guidelines and 1 Rule for Responsive Images," Cloud Four Blog, April 2, 2013. http://blog.cloudfour.com/8-guidelines-and-1-rule-for-responsive-images/.

I love the analogy of baseball for the triple play. The metaphor of creating the perfect scenario for getting all three outs: make a play that covers all of your bases and all of your customers. Now ask yourself, do you want to make a play/strategy that covers all of your own customers?

I love reading mobile anecdotes about customer usage; two-thirds of smartphone users say a mobile-friendly site makes them more likely to buy a company's product or service or 74% say they're more likely to return to the site later if it's mobile optimized.[6] What about the remaining one-third or 26%? Will they go to a mobile app or desktop site?

Why leave customers or their revenue on the table? Like the triple play, why choose a strategy that requires you to select either an app or a mobile website; why not choose both? Good user experience tells us to talk to your customers. Making the right choice of what to design for should not be based on the weather forecast or an opinion; it should be based on your own unique customer feedback. Choosing how to design your experience and for what platforms should be secondary compared to creating a mobile strategy first. Once you create a strategy, then you can fill in the details; are you a mobile app only player, are you building with RWD, or are you going to build both? Listen to your users and customers, create a plan, and then make the choice. Once you have a plan in place, then you can come up with a name for your strategy.

---

[6] Robert Hoff, "Google Research: No Mobile Site = Lost Customers," Forbes, September 25, 2012. http://www.forbes.com/sites/roberthof/2012/09/25/google-research-no-mobile-site-lost-customers/.

CHAPTER 9

The Future of Mobile UX is in Using Performance Metrics

INTRODUCTION

Let's walk through an ideal user experience project together. In the beginning, we interview and capture data from real users. After that, we build site maps, wireframes, and a prototype to gather feedback on all of those deliverables. We test these beginning structures with test users, take that feedback, and optimize our user experience design. As the user interface design is completed, we create another prototype and measure user reactions again. We gather and collect data about buttons, user impressions, how colors change conversion rates, whether images increase traffic, and more metrics. We test and collect metrics on everything about our user experience design from start to finish. With this seemingly thorough collection of data, we did not measure one of the fundamental differences of what makes the mobile different from the desktop. How fast is our mobile user experience?

> ## MOBILE MANTRA #4:
> ### SPEED IS KING!

The last mobile mantra is an important one. No user experience can be fully usable without the single most import experience design criteria: *it needs to be fast*.

We are spoiled desktop users. We use web experiences that include several video files, large images, streaming audio, and multiple pages of content—and these experiences are only getting bigger. As monitors get wider, so do our web experiences; web experiences are now being optimized for a 1200-pixel width. Apple has even introduced the retina display format for websites to make the display of images appear at a higher resolution than the traditional 72 dpi, and the file sizes for these images are even larger. Our Internet pipeline is getting larger as well, which has allowed for these experiences to be realized. It is 25 Mbps[1] one day and growing to 50 Mbps the next; by the time you read this, the Internet pipeline could be up to 100 Mbps in your home.

The creation of a desktop website is now a science. We install analytics, measure conversions (how long it takes for a user to become a paid customer), analyze online shopping trends, and build web pages to optimize to a perfect transaction or experience. If a website like Amazon.com loses one second of performance it can cost them $1.6 billion in sales over the course of a year[2]; web performance is a serious business. So much is done for desktop users behind the scenes to make the performance better, and we don't even know it. We really are spoiled desktop users who expect the world, and have it delivered, when it comes to our web experiences.

Compared to the desktop world, the mobile can be considered uncharted territory. New devices, new screen sizes, new network speeds, and new operating system versions are being introduced almost every month. It can be like a race trying to catch up with all of the changes. As the mobile matures, we have observed the desktop users' patterns of shopping, surfing, accessing data, and optimized them

---

[1] Mbps—megabits per second, a term used to measure data transfer.

[2] Kit Eaton, "How one second could cost Amazon $1.6 billion in sales." Fast Company March 15, 2012. http://www.fastcompany.com/1825005/how-one-second-could-cost-amazon-16-billion-sales.

for the mobile experience. With all of these variables in the mobile, one thing remains constant; the users' expectation of a mobile experience is that it needs to be fast. If they have just bought the phone with the fastest process on the fastest network, who will be to blame when their experience fails; it will be the app or mobile website they were last looking at.

FAILURE REDEFINED

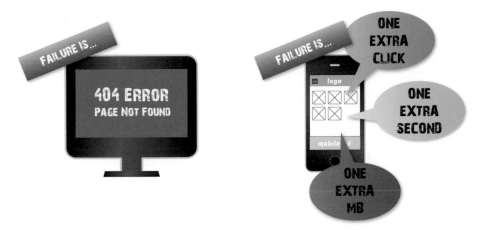

Desktop Versus Mobile Failure.

Failure to a desktop user means they get a blue screen of death (Windows) or a page failure when their web page is not loaded. For desktop users, it is a mere shrug of the shoulders when a web page fails. More than likely, they will be back surfing in no time. Desktop users are sitting at home or work for a "purpose-driven" experience. They are focused on doing tasks, buying items, signing up for a newsletter, searching the web, or reading their email. These users have some patience when they get to a roadblock.

The mobile, on the other hand, features an audience that has traditionally been more "perusal-driven," while on a bus, or in a cab, or on a train, users open their smartphones to search, tag photos, or even browse the web. Just because mobile users have much more access to the web from anywhere or any place doesn't mean they have more patience. Rather, it is the opposite. A frustrated mobile user presses the home button on their iPhone to quit the website or app they are on in less than a second; and in that second, they have left your experience frustrated. As we trend to more "purpose" driven mobile usage, a user's expectation and immediacy of having their experience load quickly becomes more critical to them. Failure to a mobile user looks different than what we as web designers are accustomed to, it is not a page loading as in the desktop, it is the mobile web or app they are on that does not load quickly enough; it is not getting to buy their items fast enough. Failure also derives from too many screens to go through to finish their task; not getting to their content or email within just a few touches. It is that extra second of loading, or that extra page in the user experience that will cost you your mobile users; this is the real failure of a mobile experience.

START DIFFERENTLY

The current trend in mobiles is to attack this performance problem at the tail end of the mobile pipeline. It is to take an experience we already have and try to optimize it to make it faster. It has become the new game in the mobile to add more JavaScript, add another CSS hack, or add expensive content acceleration to an existing mobile experience to gain ourselves half a second or two. With performance being such a critical problem in the mobile, attacking it with this approach equates to shaving off pieces of ice little by little. What we really need, is to hack a large chunk off the page's performance to see a noticeable difference. Only by starting the design of our user experience with performance in mind can we attain this goal.

How can we solve this mobile dilemma? Let's add three more mobile patterns into our set of tools. Each of these will describe common use cases where a user experience can be redesigned with performance in mind. Use these three mobile performance patterns as you start designing your mobile user experience.

Performance Pattern 1: Limiting Social Interactions

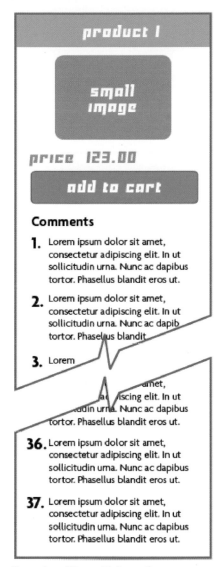

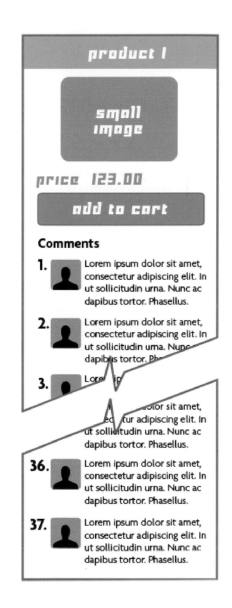

Examples of Page with Long Comments.

Being social has gone from connecting with old friends to getting every users' feedback on every topic. When building a mobile experience, it helps to have your new social family post comments, pictures, or even reviews on locations, products, or experiences. Yet there are times where being social is too social; and your mobile performance will suffer. An uncontrolled list of comments or user feedback will cause two experience problems that we will need to solve.

1. A long list of comments causes users to have to scroll too much on a page—a good user experience needs to keep scrolling to a minimum. Expect frustrated users if they need to scroll to the very bottom to read all of the comments or post their own comment. Limiting this scrolling to two screens maximum can help soften this experience. Our goal is to keep pages small and compact. Let's look at how to compact the images and text next.

2. Every comment and image will hurt you—every time you load in an image or comment, this increases the download size of your page. The more you think you are giving your users, the more you are slowing down the page. Longer pages mean longer load times. Try adding pagination to pages and displaying six to eight images or comments maximum per page. Adding a "show more" feature can help you to limit what's displayed on the screen. If you need to display hundreds, then you have too much content; archive some of this data you have collected. The goal is to keep what you load on the screen as small as possible. If a user requests a page or content they want, they are willing to wait longer than when the page first loads.

PERFORMANCE RESULTS

Average Load Times of Pattern #1

- Example of page with long comments—1.92 seconds

- Example of page with long comments (with user profile images)—3.24 seconds

- Example of page with paginated comments—1.11 seconds

See more details of this performance test in Appendix D

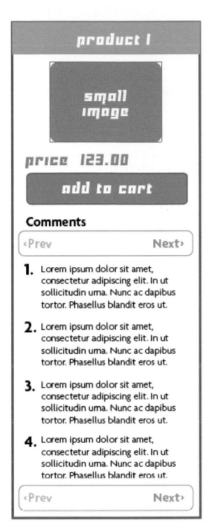

Examples of Page with Pagination.

Performance Pattern 2: Limiting Products or Images

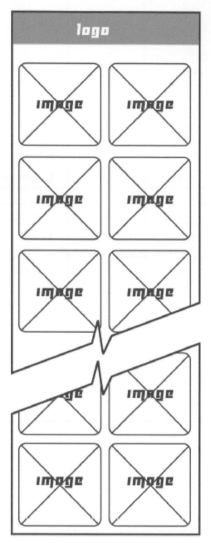

Example of Page with Multiple Images.

Like the social media page example, we need to create a limit to what we want to display to our users when we show images. Heavier than a text file, images are large, more colorful, more prominent, and can help enhance our experience,[3] but they are also our biggest performance inhibitor.

1. Every time you display images of products to sell, or friends' photographs, you are making your mobile experience slower and slower. If you are loading 40 images, you will need to contact a web server to download each one. Each request (called an http request) slows down your experience. Now imagine trying to display thirty or forty of these images on a page. This may work well on a desktop when you have the screen real estate and bandwidth, but is a sure way to frustrate your mobile user when they need to wait. In this case, we need to keep our images down to eight to ten images on each screen. We need to make each thumbnail small enough to keep, but also large enough to allow a user's thumb to touch it.

2. How can we handle more images than just eight or ten? We need to add a "Load More" button to give the users the option of what to see next. We can also give them some method of filtering up front, buying them the ability to load in the images they are expecting and want to see first.

---

[3] The Web Usability Blog, "Use pictures to direct the user's gaze," August 5, 2010. http://webusability-blog.com/use-pictures-to-direct-the-users-gaze/.

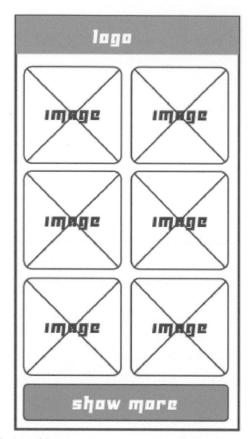

Example of Page with Paginated Images.

Performance Pattern 3: Designing Responsive Web Design (RWD) with Intent for Every Platform

480 x 800

1200 x 1024

RWD Example of Mobile Site as Copy of Desktop Website.

One of the problems with responsive web design is the performance issue of squashing and stretching a page to work at different break points. One of the main design flaws of this technique is to try to show all of the same content images into the same size frame across all widths.[4]

---

[4] Guy Podjarny, "Real World RWD Performance – Take 2," March 7, 2013. http://www.guypo.com/uncategorized/real-world-rwd-performance-take-2/.

1. Don't try to scale your images from 100 pixels wide to 400 wide. The act of doing so will cause your page to come to a screeching halt. It will work when resizing to a tablet version, but not when doing large leaps of widths. Use another set of images for a mobile version, not just what you have on the desktop version.

2. Don't bring all your text over because you can. Try to hide or limit the copy on your mobile experience. No one wants to read through large paragraphs of text on pages that scroll through eternity. Every time you load those large pages, you increase your page load time of your site, and risk losing the attention of your user, who is juggling other tasks.

3. Ask yourself the tough questions. If you have such a large experience that you want to load in on a mobile browser, is it really necessary? If designing RWD for mobile seems impossible, then why use it? No one says you can't use RWD for anything about 800 pixels wide and a mobile only version for anything lower. Our goal for the mobile is to design small and clean experiences.

480 x 800

1200 x 1024

RWD Example of Mobile Site as Unique Website.

MAKING A PLAN TO TEST ON DEVICES

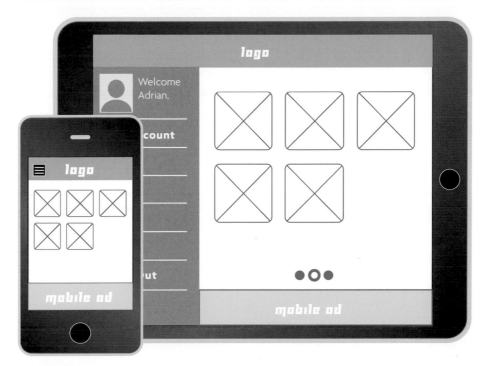

SMARTPHONE   TABLET

MOBILE MANTRA #2:
BE ON THE DEVICE!

Any good mantra is worth repeating again and again. Remember always tell yourself to start and end on a real device.

There is a big gap between creating a user experience and building it in real life. Part of this gap is filled with creating prototypes to test out our mobile experience. I am an adamant believer of testing these experiences on a real device from the real view of our users. If we are already using a device to test a prototype, why not use it to test our user experience performance as well? Start by creating a testing plan for yourself using some real content and images, then build a page you can open on several different devices. If you are designing a mobile app or mobile web page you are looking at how long your page will load on both devices. Want to see what ten images versus five images look like loading on your smartphone; test it out in a prototype of that page. Looking to see what size your images need to be on your mobile home page? Test them out in another prototype. Remember the dangers of the *prototype trap (As discussed in Chapter 7)*; focus on creating small prototypes to test a page or look and feel of an experience. The more you use these prototypes in real world scenarios, opening the page on a train, bus, or as you are walking, the more you can create performance guidelines for yourself. How has does the page load on a WiFi device versus a 3G device? How does it load on different carriers or devices? Be creative, use my exercise of visiting a carrier store to load your experience onto multiple devices. If your goal is to create a testing plan of devices for the QA of your app when it is being developed, why not use the same testing plan for measuring the performance of your user experience as well?

PROVING A POINT

User experience designers are commonly brought in to settle an argument when decisions for users have to be made. What does user experience say about breadcrumbs, what does user experience tell us about increasing conversions, or what is our opinion of a user flow, are the a few of the decisions that we are regularly asked to make. But who or what will help prove the decisions made by the user experience designer when it comes to creating a mobile experience? Performance can settle these arguments when it comes to designing user experience for mobile. Why are there only ten images instead of twenty? Why are we showing limited icons at a certain size? Why does the flow of pages work in a certain manner? I am a big believer that mobile user experience is about having a *point of view*. That point of view can be even more powerful when you have performance metrics to base your decisions on.

A CALL TO ACTION

Having a network of friends and family to test your experiences or a bag full of smartphones at your house can only take you so far. What we need is a large network of devices that we can use to test our mobile user experience on. If it's part of your social family, a co-op of shared devices, or a potential product; the future is in sharing the testing of our user experiences across the country. Our pages need to be smarter, faster, and to travel and be tested across borders. The mobile has exposed our user experiences internationally, and at scales we cannot imagine; as mobile usage explodes so will the exposure of the experiences we create. Forget about getting traffic in the thousands; be prepared to have your mobile experience visited by hundreds of thousands or millions of users. If this is the case, why would you test your mobile experience on a single phone in your drawer?

List of Devices from Chapter 1

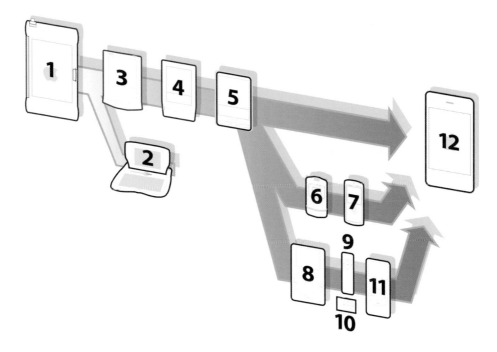

DEVICE LIST

PDAS

1. Apple Newton MessagePad 2000, 1998

2. Apple eMate 300, 1997

3. Palm V, 1999

4. Palm IIIc, 2000

5. Sony Clie, 2001

FEATURE PHONES

6. Motorola Razr, 2004

7. Motorola Slvr, 2005

MP3 PLAYERS

8. Apple iPod (First Generation), 2001

9. Apple iPod Shuffle (First Generation), 2005

10. Apple iPod Shuffle (Second Generation), 2006

11. Apple iPod Nano (First Generation), 2005

SMARTPHONES

12. Apple iPhone (First Generation), 2007

List of My Devices from Chapter 4

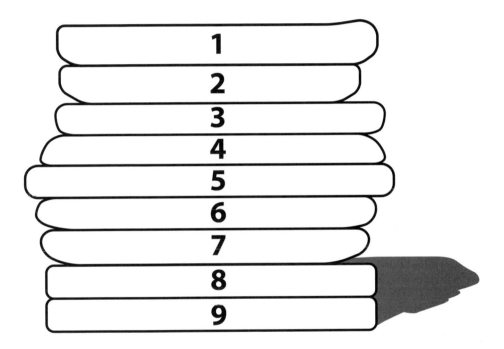

DEVICE LIST
PICTURED

1. Google ION (Unlocked), Android 1.6, screen size: 320×480 pixels.

2. HTC Aria (AT&T), Android 2.3, screen size: 320×480 pixels.

3. Kyocera Event (Virgin Mobile), Android 4.1, screen size: 320×480 pixels.

4. Samsung Captivate (AT&T), Android 2.4, screen size: 480×800 pixels.

5. HTC One S (T-Mobile), Android 4.1, screen size: 540×960 pixels.

6. Nokia Lumia 800 (Unlocked), Windows Phone 7.5, screen size: 480×800 pixels.

7. Apple iPhone 3GS (AT&T), iOS 6.0.1, screen size: 320×480 pixels.

8. Apple iPhone 4 (AT&T), iOS 6.0.3, screen size: 640×960 pixels.

9. Apple iPhone 4S (Sprint), iOS 5.1.1, screen size: 640×960 pixels.

NOT PICTURED

10. Apple iPad (Wi-fi), iOS 5.1.1, screen size: 1024×768 pixels.

11. Motorola Xyboard 10.1 (Verizon), Android 4.1, screen size: 1280×800 pixels.

12. Apple iPhone 5 (AT&T), iOS 6.0.3, screen size: 640×1136 pixels.

MUSICAL SIM CARDS

You may have noticed, that I have quite a few mobile devices. A little secret, not all of my phones are on a 2-year contract for a mobile plan. In fact, I only have three that are. Here are two tips to get you up and running on real devices.

1. I have purchased and borrowed several phones on AT&T. As they are all on the same carrier I can switch my SIM card between all phones. With the help of a few SIM card adapters I can switch between phones with mini-, micro-, and regular-sized SIM card slots. The voice and data plan works on all of the devices.

2. Instead of looking at devices with a yearly contract look at noncontract carriers like Virgin, Boost, and T-Mobile. Most noncontract carriers provide smartphones and data plans, even the iPhone.

Sample Mobile Sheets from Chapters 5 and 7

Use these templates to wireframe, prototype, or sketch out quick ideas. Here are a few easy steps describing how to use them.

STEP 1: CUT MOBILE SHEETS OUT OF THIS BOOK

STEP 2: SCAN THE PAGES OR PHOTOCOPY THEM

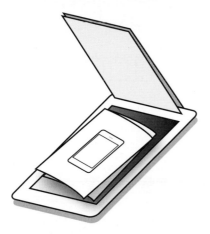

STEP 3: DRAW ON THEM!

Name _____   Date _____

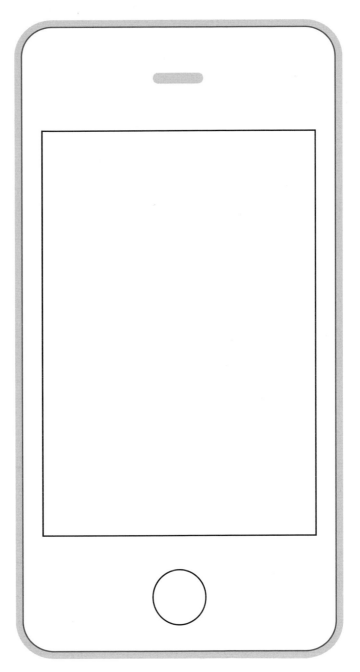

iPhone 4

Project _____

Version _____

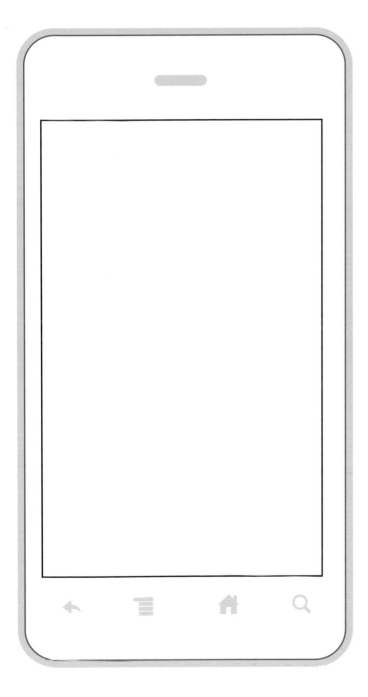

Android 2.X

Project _____

Version _____

Android 4.X

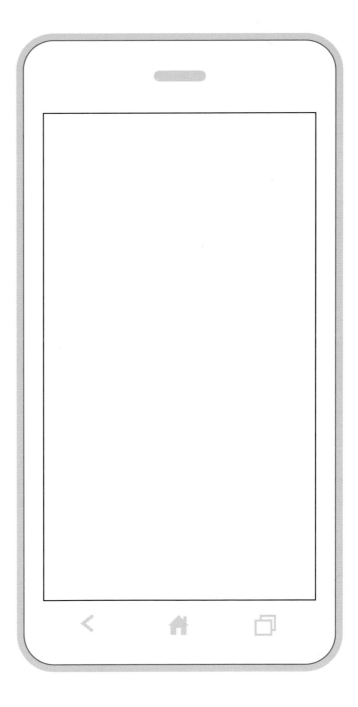

Name _____   Date _____

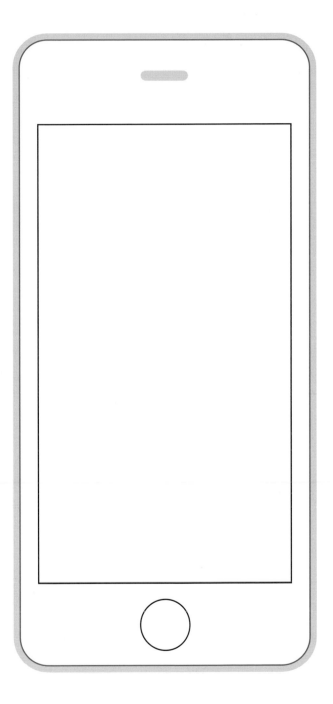

iPhone 5

Project _____

Version _____

iPhone 4

Name _____

Date _____

Project _____

Version _____

Android 2.X

Name _____

Date _____

Project _____ Version _____

Android 4.X

Name _____

Date _____

Project _____

Version _____

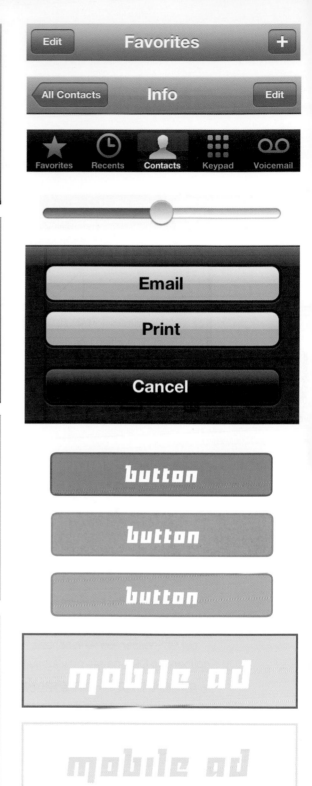

Q W E R T Y U I O P
A S D F G H J K L
Z X C V B N M
@123 . / .com Go

Q W E R T Y U I O P
A S D F G H J K L
Z X C V B N M
123 space Go

1 2 3 4 5 6 7 8 9 0
- / : ; ( ) $ & @ "
#+= . , ? ! '
ABC space return

1   2 ABC   3 DEF
4 GHI   5 JKL   6 MNO
7 PQRS   8 TUV   9 WXYZ
+*#   0

Edit   Favorites   +

All Contacts   Info   Edit

Favorites   Recents   Contacts   Keypad   Voicemail

Email

Print

Cancel

button

button

button

mobile ad

mobile ad

UI Elements

iOS 7

All Contacts   Edit

Favorites   Recents   Contacts   Keypad   Voicemail

Delete Note

Cancel

button

button

button

mobile ad

mobile ad

Q	W	E	R	T	Y	U	I	O	P
A	S	D	F	G	H	J	K	L
Z	X	C	V	B	N	M
123	space	Go

Q	W	E	R	T	Y	U	I	O	P
A	S	D	F	G	H	J	K	L
Z	X	C	V	B	N	M
123	space	Go

1	2	3	4	5	6	7	8	9	0
-	/	:	;	(	)	$	&	@	"
#+=	.	,	?	!	'
ABC	space	return

1	2 ABC	3 DEF
4 GHI	5 JKL	6 MNO
7 PQRS	8 TUV	9 WXYZ
#+*	0	⌫

iOS 6

Edit   Favorites   +

All Contacts   Info   Edit

Favorites   Recents   Contacts   Keypad   Voicemail

Email

Print

Cancel

button

button

button

mobile ad

mobile ad

Q	W	E	R	T	Y	U	I	O	P
A	S	D	F	G	H	J	K	L
Z	X	C	V	B	N	M
@123	.	/	.com	Go

Q	W	E	R	T	Y	U	I	O	P
A	S	D	F	G	H	J	K	L
Z	X	C	V	B	N	M
123	space	Go

1	2	3	4	5	6	7	8	9	0
-	/	:	;	(	)	$	&	@	"
#+=	.	,	?	!	'
ABC	space	return

1	2 ABC	3 DEF
4 GHI	5 JKL	6 MNO
7 PQRS	8 TUV	9 WXYZ
#+*	0	⌫

Mobile Performance Results from Chapter 9

TESTING METHODOLOGY

These mobile performance tests were created using the Marlin Mobile Performance Monitoring tool. Here are the characteristics of this test.

- Created pages coded in HTML5 for each pattern example

- Used three Android devices (smartphones and tablets)

- Ran the test 12 times a day

- Ran the test over a period of 12 days

- All devices were connected to a mobile carrier

- Pages hosted on the same server

- Did not remove outliers

PERFORMANCE PATTERN #1
URLs TESTED

1. Example of page with long comments

   http://mobileuxbook.com/pattern1/longcomments.html

2. Example of page with long comments (with user profile images)

   http://mobileuxbook.com/pattern1/longcomments2.html

3. Example of page with paginated comments

   http://mobileuxbook.com/pattern1/shortcomments.html

RESULTS
Average Page Load Time (Seconds): (*from High to Low*)

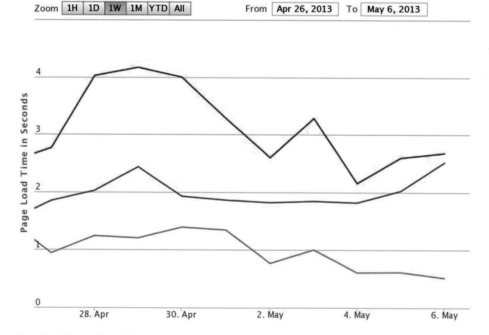

Page Load Time (Seconds), Source Marlin Mobile.

Mobile User Experience | Patterns to Make Sense of it All

	URLs Tested	Average page load time (s)	Page weight (KB)
1.	Long comments (with user profile images)	3.24	206
2.	Long comments	1.92	38.2
3.	Paginated comments	1.11	31.8

Average Page Load Time by Carrier (Seconds), *(from High to Low)*

	URLs Tested	Verizon	T-Mobile	Virgin Mobile
1.	Long comments (with user profile images)	3.4	3.03	3.33
2.	Long comments	1.91	2.12	1.8
3.	Paginated comments	1.25	0.72	1.32

Average Page Load Time by Device (Seconds), *(from High to Low)*

	URLs Tested	Motorola Xyboard 10.1 (Tablet)	HTC One S (Smartphone)	Kyocera Event (Smartphone)
1.	Long comments (with user profile images)	3.4	3.03	3.33
2.	Long comments	1.91	2.12	1.8
3.	Paginated comments	1.25	0.72	1.32

PERFORMANCE PATTERN #2
URLS TESTED

1. Example of page with multiple images—4.3 s

   http://mobileuxbook.com/pattern2/longimages.html

2. Example of page with paginated images—1.51 s

   http://mobileuxbook.com/pattern2/shortimages.html

RESULTS
Average Page Load Time (Seconds), (*from High to Low*)

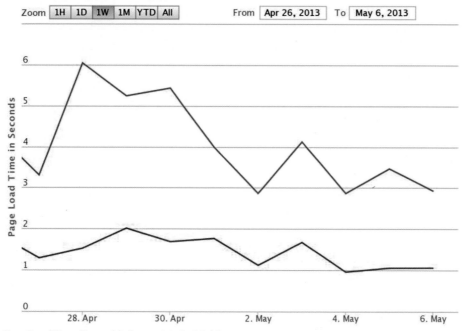

Page Load Time (Seconds), Source Marlin Mobile.

	URLs Tested	Average page load time (s)	Page weight (KB)
1.	Multiple images	4.3	1,000
2.	Paginated images	1.51	209

Average Page Load Time by Carrier (Seconds), (*from High to Low*)

	URLs Tested	Verizon	T-Mobile	Virgin Mobile
1.	Multiple images	6.78	4.36	3.32
2.	Paginated images	1.85	1.24	1.57

Average Page Load Time by Device (Seconds), (*from High to Low*)

	URLs Tested	Motorola Xyboard 10.1 (Tablet)	HTC One S (Smartphone)	Kyocera Event (Smartphone)
1.	Multiple images	6.78	4.36	3.32
2.	Paginated images	1.85	1.24	1.57

Want to Learn More about Mobile Performance Monitoring on Real Devices?

Visit www.marlinmobile.com to learn more about how you can test your mobile web pages on real devices.

Index

Note: Page numbers followed by *b* indicate boxes and *f* indicate figures.